MW01254845

SAXON MATH™

Course 2

Stephen Hake

Power-Up Workbook

SAXON™
A Harcourt Achieve Imprint
www.SaxonPublishers.com
1-800-284-7019

Printed in the U.S.A.

ISBN 978-1-591-41873-3

25 26 27 28 29 30 0982 22

4500849470 A B C D E F G

Dear Student,

We enjoy watching the adventures of "Super Heroes" because they have power and they use their powers for good. Power is the ability to get things done. We acquire power through concentrated effort and practice. We build powerful bodies with vigorous exercise and healthy living. We develop powerful minds by learning and using skills that help us understand the world around us and solve problems that come our way.

We can build our mathematical power several ways. We can use our memory to store and instantly recall frequently used information. We can improve our ability to solve many kinds of problems mentally without using pencil and paper or a calculator. We can also expand the range of strategies we use to approach and solve new problems.

The Power Up section of each lesson in *Saxon Math Course 2* is designed to build your mathematical power. Each Power Up has three parts, Facts Practice, Mental Math, and Problem Solving. The three parts are printed on every Power Up page where you will record your answers. This workbook contains a Power Up page for every lesson.

Facts Practice is like a race—write the answers as fast as you can without making mistakes. If the information in the Fact Practice is new to you, take time to study the information so that you can recall the facts quickly and can complete the exercise faster next time.

Mental Math is the ability to work with numbers in your head. This skill greatly improves with practice. Each lesson includes several mental math problems. Your teacher will read these to you or ask you to read them in your book. Do your best to find the answer to each problem without using pencil and paper, except to record your answers. Strong mental math ability will help you throughout your life.

Problem Solving is like a puzzle. You need to figure out how to solve the puzzle. There are many different strategies you can use to solve problems. There are also some questions you can ask yourself to better understand a problem and come up with a plan to solve it. Your teacher will guide you through the problem each day. Becoming a good problem solver is a superior skill that is highly rewarded.

The Power Ups will help you excel at math and acquire math power that will serve you well for the rest of you life.

Stephen Hake

Temple City, California

Power-Up Workbook

Name ______________________________

Student Progress Chart

Fact Practice

Power Up Facts	# Possible	Time and Score (time / # correct)									
A 40 Multiplication Facts	40										
B 20 Equations	20										
C 20 Improper Fractions and Mixed Numbers	20										
D 20 Fractions to Reduce	20										
E Circles	12										
F Lines, Angles, Polygons	12										
G + − × ÷ Fractions	16										
H Measurement Facts	30										
I Proportions	15										
J + − × ÷ Decimals	16										
K Powers and Roots	20										
L Fraction-Decimal-Percent Equivalents	24										
M Metric Conversions	22										
N + − × ÷ Mixed Numbers	16										
O Classifying Quadrilaterals and Triangles	8										
P + − × ÷ Integers	16										
Q Percent-Decimal-Fraction Equivalents	24										
R Area	8										
S Scientific Notation	12										
T Order of Operations	8										
U Two-Step Equations	12										
V + − × ÷ Algebraic Terms	16										
W Multiplying and Dividing in Scientific Notation	12										

Name ______________________ Time __________

Power Up A

Use with Lesson 1

Facts Multiply.

9×8	8×2	10×10	6×3	4×2	5×5	9×9	6×4	9×6	7×3
9×3	6×5	0×0	7×6	8×8	7×4	5×3	9×7	2×2	8×6
7×7	6×2	4×3	8×5	4×4	3×2	$n \times 0$	8×4	6×6	9×2
8×3	5×4	$n \times 1$	7×2	9×5	8×7	3×3	9×4	5×2	7×5

Mental Math

a.	**b.**	**c.**	**d.**
e.	**f.**	**g.**	**h.**

Problem Solving

Understand
What information am I given?
What am I asked to find or do?

Plan
How can I use the information I am given?
Which strategy should I try?

Solve
Did I follow the plan?
Did I show my work?
Did I write the answer?

Check
Did I use the correct information?
Did I do what was asked?
Is my answer reasonable?

Name ______________________ Time __________

Power Up A

Use with Lesson 2

Facts

Multiply.

9×8	8×2	10×10	6×3	4×2	5×5	9×9	6×4	9×6	7×3
9×3	6×5	0×0	7×6	8×8	7×4	5×3	9×7	2×2	8×6
7×7	6×2	4×3	8×5	4×4	3×2	$n \times 0$	8×4	6×6	9×2
8×3	5×4	$n \times 1$	7×2	9×5	8×7	3×3	9×4	5×2	7×5

Mental Math

a.	**b.**	**c.**	**d.**
e.	**f.**	**g.**	**h.**

Problem Solving

Understand
What information am I given?
What am I asked to find or do?

Plan
How can I use the information I am given?
Which strategy should I try?

Solve
Did I follow the plan?
Did I show my work?
Did I write the answer?

Check
Did I use the correct information?
Did I do what was asked?
Is my answer reasonable?

Name ______________________ Time ____________

Power Up A

Use with Lesson 3

Facts

Multiply.

9×8	8×2	10×10	6×3	4×2	5×5	9×9	6×4	9×6	7×3
9×3	6×5	0×0	7×6	8×8	7×4	5×3	9×7	2×2	8×6
7×7	6×2	4×3	8×5	4×4	3×2	$n \times 0$	8×4	6×6	9×2
8×3	5×4	$n \times 1$	7×2	9×5	8×7	3×3	9×4	5×2	7×5

Mental Math

a.	**b.**	**c.**	**d.**
e.	**f.**	**g.**	**h.**

Problem Solving

Understand
What information am I given?
What am I asked to find or do?

Plan
How can I use the information I am given?
Which strategy should I try?

Solve
Did I follow the plan?
Did I show my work?
Did I write the answer?

Check
Did I use the correct information?
Did I do what was asked?
Is my answer reasonable?

Name ______________________ Time __________

Power Up A

Use with Lesson 4

Facts Multiply.

9×8	8×2	10×10	6×3	4×2	5×5	9×9	6×4	9×6	7×3
9×3	6×5	0×0	7×6	8×8	7×4	5×3	9×7	2×2	8×6
7×7	6×2	4×3	8×5	4×4	3×2	$n \times 0$	8×4	6×6	9×2
8×3	5×4	$n \times 1$	7×2	9×5	8×7	3×3	9×4	5×2	7×5

Mental Math

a.	**b.**	**c.**	**d.**
e.	**f.**	**g.**	**h.**

Problem Solving

Understand
What information am I given?
What am I asked to find or do?

Plan
How can I use the information I am given?
Which strategy should I try?

Solve
Did I follow the plan?
Did I show my work?
Did I write the answer?

Check
Did I use the correct information?
Did I do what was asked?
Is my answer reasonable?

Name ____________________ Time ____________

Power Up A

Use with Lesson 5

Facts Multiply.

9×8	8×2	10×10	6×3	4×2	5×5	9×9	6×4	9×6	7×3
9×3	6×5	0×0	7×6	8×8	7×4	5×3	9×7	2×2	8×6
7×7	6×2	4×3	8×5	4×4	3×2	$n \times 0$	8×4	6×6	9×2
8×3	5×4	$n \times 1$	7×2	9×5	8×7	3×3	9×4	5×2	7×5

Mental Math

a.	**b.**	**c.**	**d.**
e.	**f.**	**g.**	**h.**

Problem Solving

Understand
What information am I given?
What am I asked to find or do?

Plan
How can I use the information I am given?
Which strategy should I try?

Solve
Did I follow the plan?
Did I show my work?
Did I write the answer?

Check
Did I use the correct information?
Did I do what was asked?
Is my answer reasonable?

Name ______________________ Time ____________

Power Up B

Use with Lesson 6

Facts

Solve each equation.

$a + 12 = 20$ $a =$	$b - 8 = 10$ $b =$	$5c = 40$ $c =$	$\frac{d}{4} = 12$ $d =$	$11 + e = 24$ $e =$
$25 - f = 10$ $f =$	$10g = 60$ $g =$	$\frac{24}{h} = 6$ $h =$	$15 = j + 8$ $j =$	$20 = k - 5$ $k =$
$30 = 6m$ $m =$	$9 = \frac{n}{3}$ $n =$	$18 = 6 + p$ $p =$	$5 = 15 - q$ $q =$	$36 = 4r$ $r =$
$2 = \frac{16}{s}$ $s =$	$t + 8 = 12$ $t =$	$u - 15 = 30$ $u =$	$8v = 48$ $v =$	$\frac{w}{3} = 6$ $w =$

Mental Math

a.	**b.**	**c.**	**d.**
e.	**f.**	**g.**	**h.**

Problem Solving

Understand
What information am I given?
What am I asked to find or do?

Plan
How can I use the information I am given?
Which strategy should I try?

Solve
Did I follow the plan?
Did I show my work?
Did I write the answer?

Check
Did I use the correct information?
Did I do what was asked?
Is my answer reasonable?

Name ______________________ Time ____________

Power Up B

Use with Lesson 7

Facts

Solve each equation.

$a + 12 = 20$ $a =$	$b - 8 = 10$ $b =$	$5c = 40$ $c =$	$\frac{d}{4} = 12$ $d =$	$11 + e = 24$ $e =$
$25 - f = 10$ $f =$	$10g = 60$ $g =$	$\frac{24}{h} = 6$ $h =$	$15 = j + 8$ $j =$	$20 = k - 5$ $k =$
$30 = 6m$ $m =$	$9 = \frac{n}{3}$ $n =$	$18 = 6 + p$ $p =$	$5 = 15 - q$ $q =$	$36 = 4r$ $r =$
$2 = \frac{16}{s}$ $s =$	$t + 8 = 12$ $t =$	$u - 15 = 30$ $u =$	$8v = 48$ $v =$	$\frac{w}{3} = 6$ $w =$

Mental Math

a.	**b.**	**c.**	**d.**
e.	**f.**	**g.**	**h.**

Problem Solving

Understand
What information am I given?
What am I asked to find or do?

Plan
How can I use the information I am given?
Which strategy should I try?

Solve
Did I follow the plan?
Did I show my work?
Did I write the answer?

Check
Did I use the correct information?
Did I do what was asked?
Is my answer reasonable?

Name ______________________ Time ____________

Power Up A

Use with Lesson 8

Facts

Multiply.

9×8	8×2	10×10	6×3	4×2	5×5	9×9	6×4	9×6	7×3
9×3	6×5	0×0	7×6	8×8	7×4	5×3	9×7	2×2	8×6
7×7	6×2	4×3	8×5	4×4	3×2	$n \times 0$	8×4	6×6	9×2
8×3	5×4	$n \times 1$	7×2	9×5	8×7	3×3	9×4	5×2	7×5

Mental Math

a.	**b.**	**c.**	**d.**
e.	**f.**	**g.**	**h.**

Problem Solving

Understand

What information am I given?

What am I asked to find or do?

Plan

How can I use the information I am given?

Which strategy should I try?

Solve

Did I follow the plan?

Did I show my work?

Did I write the answer?

Check

Did I use the correct information?

Did I do what was asked?

Is my answer reasonable?

Name ______________________ Time ____________

Power Up A

Use with Lesson 9

Facts Multiply.

9×8	8×2	10×10	6×3	4×2	5×5	9×9	6×4	9×6	7×3
9×3	6×5	0×0	7×6	8×8	7×4	5×3	9×7	2×2	8×6
7×7	6×2	4×3	8×5	4×4	3×2	$n \times 0$	8×4	6×6	9×2
8×3	5×4	$n \times 1$	7×2	9×5	8×7	3×3	9×4	5×2	7×5

Mental Math

a.	**b.**	**c.**	**d.**
e.	**f.**	**g.**	**h.**

Problem Solving

Understand
What information am I given?
What am I asked to find or do?

Plan
How can I use the information I am given?
Which strategy should I try?

Solve
Did I follow the plan?
Did I show my work?
Did I write the answer?

Check
Did I use the correct information?
Did I do what was asked?
Is my answer reasonable?

Name ______________________ Time ____________

Power Up A

Use with Lesson 10

Facts Multiply.

9×8	8×2	10×10	6×3	4×2	5×5	9×9	6×4	9×6	7×3
9×3	6×5	0×0	7×6	8×8	7×4	5×3	9×7	2×2	8×6
7×7	6×2	4×3	8×5	4×4	3×2	$n \times 0$	8×4	6×6	9×2
8×3	5×4	$n \times 1$	7×2	9×5	8×7	3×3	9×4	5×2	7×5

Mental Math

a.	b.	c.	d.
e.	f.	g.	h.

Problem Solving

Understand
What information am I given?
What am I asked to find or do?

Plan
How can I use the information I am given?
Which strategy should I try?

Solve
Did I follow the plan?
Did I show my work?
Did I write the answer?

Check
Did I use the correct information?
Did I do what was asked?
Is my answer reasonable?

Facts

Write each improper fraction as a whole number or mixed number.

$\frac{5}{2} =$	$\frac{7}{4} =$	$\frac{12}{5} =$	$\frac{10}{3} =$	$\frac{15}{2} =$
$\frac{15}{5} =$	$\frac{11}{8} =$	$2\frac{3}{2} =$	$4\frac{5}{4} =$	$3\frac{7}{4} =$

Write each mixed number as an improper fraction.

$1\frac{1}{2} =$	$2\frac{2}{3} =$	$3\frac{3}{4} =$	$2\frac{1}{2} =$	$6\frac{2}{3} =$
$2\frac{3}{4} =$	$3\frac{1}{3} =$	$4\frac{1}{2} =$	$1\frac{7}{8} =$	$12\frac{1}{2} =$

Mental Math

a.	**b.**	**c.**	**d.**
e.	**f.**	**g.**	**h.**

Problem Solving

Understand

What information am I given?

What am I asked to find or do?

Plan

How can I use the information I am given?

Which strategy should I try?

Solve

Did I follow the plan?

Did I show my work?

Did I write the answer?

Check

Did I use the correct information?

Did I do what was asked?

Is my answer reasonable?

Name ____________________ Time __________

Power Up C

Use with Lesson 12

Facts

Write each improper fraction as a whole number or mixed number.

$\frac{5}{2} =$	$\frac{7}{4} =$	$\frac{12}{5} =$	$\frac{10}{3} =$	$\frac{15}{2} =$
$\frac{15}{5} =$	$\frac{11}{8} =$	$2\frac{3}{2} =$	$4\frac{5}{4} =$	$3\frac{7}{4} =$

Write each mixed number as an improper fraction.

$1\frac{1}{2} =$	$2\frac{2}{3} =$	$3\frac{3}{4} =$	$2\frac{1}{2} =$	$6\frac{2}{3} =$
$2\frac{3}{4} =$	$3\frac{1}{3} =$	$4\frac{1}{2} =$	$1\frac{7}{8} =$	$12\frac{1}{2} =$

Mental Math

a.	**b.**	**c.**	**d.**
e.	**f.**	**g.**	**h.**

Problem Solving

Understand

What information am I given?

What am I asked to find or do?

Plan

How can I use the information I am given?

Which strategy should I try?

Solve

Did I follow the plan?

Did I show my work?

Did I write the answer?

Check

Did I use the correct information?

Did I do what was asked?

Is my answer reasonable?

Name ______________________ Time __________

Power Up C

Use with Lesson 13

Facts

Write each improper fraction as a whole number or mixed number.

$\frac{5}{2} =$	$\frac{7}{4} =$	$\frac{12}{5} =$	$\frac{10}{3} =$	$\frac{15}{2} =$
$\frac{15}{5} =$	$\frac{11}{8} =$	$2\frac{3}{2} =$	$4\frac{5}{4} =$	$3\frac{7}{4} =$

Write each mixed number as an improper fraction.

$1\frac{1}{2} =$	$2\frac{2}{3} =$	$3\frac{3}{4} =$	$2\frac{1}{2} =$	$6\frac{2}{3} =$
$2\frac{3}{4} =$	$3\frac{1}{3} =$	$4\frac{1}{2} =$	$1\frac{7}{8} =$	$12\frac{1}{2} =$

Mental Math

a.	**b.**	**c.**	**d.**
e.	**f.**	**g.**	**h.**

Problem Solving

Understand

What information am I given?

What am I asked to find or do?

Plan

How can I use the information I am given?

Which strategy should I try?

Solve

Did I follow the plan?

Did I show my work?

Did I write the answer?

Check

Did I use the correct information?

Did I do what was asked?

Is my answer reasonable?

Name ______________________ Time __________

Power Up A

Use with Lesson 14

Facts

Multiply.

9×8	8×2	10×10	6×3	4×2	5×5	9×9	6×4	9×6	7×3
9×3	6×5	0×0	7×6	8×8	7×4	5×3	9×7	2×2	8×6
7×7	6×2	4×3	8×5	4×4	3×2	$n \times 0$	8×4	6×6	9×2
8×3	5×4	$n \times 1$	7×2	9×5	8×7	3×3	9×4	5×2	7×5

Mental Math

a.	**b.**	**c.**	**d.**
e.	**f.**	**g.**	**h.**

Problem Solving

Understand

What information am I given?

What am I asked to find or do?

Plan

How can I use the information I am given?

Which strategy should I try?

Solve

Did I follow the plan?

Did I show my work?

Did I write the answer?

Check

Did I use the correct information?

Did I do what was asked?

Is my answer reasonable?

Name ______________________ Time ___________

Power Up C

Use with Lesson 15

Facts

Write each improper fraction as a whole number or mixed number.

$\frac{5}{2}$ =	$\frac{7}{4}$ =	$\frac{12}{5}$ =	$\frac{10}{3}$ =	$\frac{15}{2}$ =
$\frac{15}{5}$ =	$\frac{11}{8}$ =	$2\frac{3}{2}$ =	$4\frac{5}{4}$ =	$3\frac{7}{4}$ =

Write each mixed number as an improper fraction.

$1\frac{1}{2}$ =	$2\frac{2}{3}$ =	$3\frac{3}{4}$ =	$2\frac{1}{2}$ =	$6\frac{2}{3}$ =
$2\frac{3}{4}$ =	$3\frac{1}{3}$ =	$4\frac{1}{2}$ =	$1\frac{7}{8}$ =	$12\frac{1}{2}$ =

Mental Math

a.	**b.**	**c.**	**d.**
e.	**f.**	**g.**	**h.**

Problem Solving

Understand

What information am I given?

What am I asked to find or do?

Plan

How can I use the information I am given?

Which strategy should I try?

Solve

Did I follow the plan?

Did I show my work?

Did I write the answer?

Check

Did I use the correct information?

Did I do what was asked?

Is my answer reasonable?

Name ______________________ Time __________

Power Up D

Use with Lesson 16

Facts Reduce each fraction to lowest terms.

$\frac{50}{100} =$	$\frac{4}{16} =$	$\frac{6}{8} =$	$\frac{8}{12} =$	$\frac{10}{100} =$
$\frac{8}{16} =$	$\frac{20}{100} =$	$\frac{3}{12} =$	$\frac{60}{100} =$	$\frac{9}{12} =$
$\frac{6}{9} =$	$\frac{90}{100} =$	$\frac{5}{10} =$	$\frac{12}{16} =$	$\frac{25}{100} =$
$\frac{4}{10} =$	$\frac{4}{6} =$	$\frac{75}{100} =$	$\frac{4}{12} =$	$\frac{6}{10} =$

Mental Math

a.	b.	c.	d.
e.	f.	g.	h.

Problem Solving

Understand

What information am I given?

What am I asked to find or do?

Plan

How can I use the information I am given?

Which strategy should I try?

Solve

Did I follow the plan?

Did I show my work?

Did I write the answer?

Check

Did I use the correct information?

Did I do what was asked?

Is my answer reasonable?

Name ______________________ Time __________

Power Up A

Use with Lesson 17

Facts Multiply.

9 × 8	8 × 2	10 × 10	6 × 3	4 × 2	5 × 5	9 × 9	6 × 4	9 × 6	7 × 3
9 × 3	6 × 5	0 × 0	7 × 6	8 × 8	7 × 4	5 × 3	9 × 7	2 × 2	8 × 6
7 × 7	6 × 2	4 × 3	8 × 5	4 × 4	3 × 2	n × 0	8 × 4	6 × 6	9 × 2
8 × 3	5 × 4	n × 1	7 × 2	9 × 5	8 × 7	3 × 3	9 × 4	5 × 2	7 × 5

Mental Math

a.	**b.**	**c.**	**d.**
e.	**f.**	**g.**	**h.**

Problem Solving

Understand

What information am I given?

What am I asked to find or do?

Plan

How can I use the information I am given?

Which strategy should I try?

Solve

Did I follow the plan?

Did I show my work?

Did I write the answer?

Check

Did I use the correct information?

Did I do what was asked?

Is my answer reasonable?

Name ______________________ Time __________

Power Up D

Use with Lesson 18

Facts

Reduce each fraction to lowest terms.

$\frac{50}{100} =$	$\frac{4}{16} =$	$\frac{6}{8} =$	$\frac{8}{12} =$	$\frac{10}{100} =$
$\frac{8}{16} =$	$\frac{20}{100} =$	$\frac{3}{12} =$	$\frac{60}{100} =$	$\frac{9}{12} =$
$\frac{6}{9} =$	$\frac{90}{100} =$	$\frac{5}{10} =$	$\frac{12}{16} =$	$\frac{25}{100} =$
$\frac{4}{10} =$	$\frac{4}{6} =$	$\frac{75}{100} =$	$\frac{4}{12} =$	$\frac{6}{10} =$

Mental Math

a.	**b.**	**c.**	**d.**
e.	**f.**	**g.**	**h.**

Problem Solving

Understand

What information am I given?

What am I asked to find or do?

Plan

How can I use the information I am given?

Which strategy should I try?

Solve

Did I follow the plan?

Did I show my work?

Did I write the answer?

Check

Did I use the correct information?

Did I do what was asked?

Is my answer reasonable?

Name ______________________ Time ____________

Power Up C

Use with Lesson 19

Facts Write each improper fraction as a whole number or mixed number.

$\frac{5}{2} =$	$\frac{7}{4} =$	$\frac{12}{5} =$	$\frac{10}{3} =$	$\frac{15}{2} =$
$\frac{15}{5} =$	$\frac{11}{8} =$	$2\frac{3}{2} =$	$4\frac{5}{4} =$	$3\frac{7}{4} =$

Write each mixed number as an improper fraction.

$1\frac{1}{2} =$	$2\frac{2}{3} =$	$3\frac{3}{4} =$	$2\frac{1}{2} =$	$6\frac{2}{3} =$
$2\frac{3}{4} =$	$3\frac{1}{3} =$	$4\frac{1}{2} =$	$1\frac{7}{8} =$	$12\frac{1}{2} =$

Mental Math

a.	**b.**	**c.**	**d.**
e.	**f.**	**g.**	**h.**

Problem Solving

Understand

What information am I given?

What am I asked to find or do?

Plan

How can I use the information I am given?

Which strategy should I try?

Solve

Did I follow the plan?

Did I show my work?

Did I write the answer?

Check

Did I use the correct information?

Did I do what was asked?

Is my answer reasonable?

Name ______________________________ Time ____________

Power Up D

Use with Lesson 20

Facts Reduce each fraction to lowest terms.

$\frac{50}{100} =$	$\frac{4}{16} =$	$\frac{6}{8} =$	$\frac{8}{12} =$	$\frac{10}{100} =$
$\frac{8}{16} =$	$\frac{20}{100} =$	$\frac{3}{12} =$	$\frac{60}{100} =$	$\frac{9}{12} =$
$\frac{6}{9} =$	$\frac{90}{100} =$	$\frac{5}{10} =$	$\frac{12}{16} =$	$\frac{25}{100} =$
$\frac{4}{10} =$	$\frac{4}{6} =$	$\frac{75}{100} =$	$\frac{4}{12} =$	$\frac{6}{10} =$

Mental Math

a.	**b.**	**c.**	**d.**
e.	**f.**	**g.**	**h.**

Problem Solving

Understand

What information am I given?

What am I asked to find or do?

Plan

How can I use the information I am given?

Which strategy should I try?

Solve

Did I follow the plan?

Did I show my work?

Did I write the answer?

Check

Did I use the correct information?

Did I do what was asked?

Is my answer reasonable?

Name ______________________ Time __________

Power Up E

Use with Lesson 21

Facts

Write the word or words to complete each definition.

The distance around a circle is its  ________.	Every point on a circle is the same distance from its 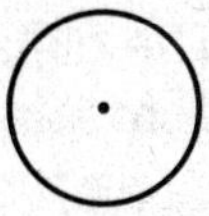 ________.	The distance across a circle through its center is its 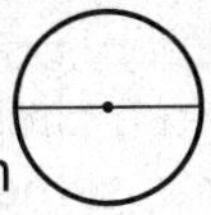 ________.	The distance from a circle to its center is its 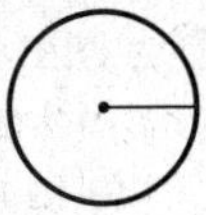 ________.
Two or more circles with the same center are 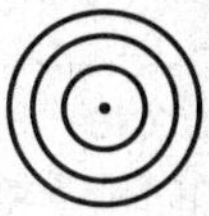 ________.	A segment between two points on a circle is a 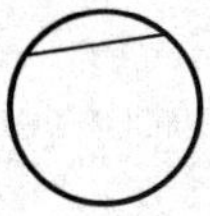 ________.	Part of a circumference is an  ________.	Part of a circle bounded by an arc and two radii is a 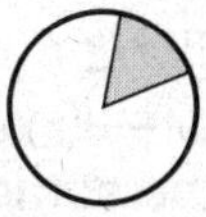 ________.
Half a circle is a  ________.	An angle whose vertex is the center of a circle is a 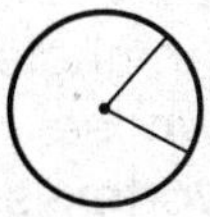 ________.	An angle whose vertex is on the circle whose sides include chords is an 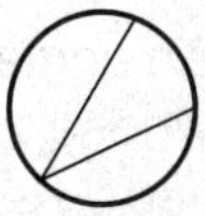 ________.	A polygon whose vertices are on the circle and whose edges are within the circle is an 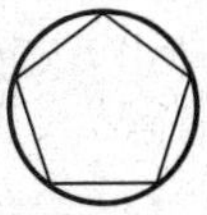 ________.

Mental Math

a.	**b.**	**c.**	**d.**
e.	**f.**	**g.**	**h.**

Problem Solving

Understand

What information am I given?

What am I asked to find or do?

Plan

How can I use the information I am given?

Which strategy should I try?

Solve

Did I follow the plan?

Did I show my work?

Did I write the answer?

Check

Did I use the correct information?

Did I do what was asked?

Is my answer reasonable?

Name ____________________ Time __________

Power Up E

Use with Lesson 22

Facts

Write the word or words to complete each definition.

The distance around a circle is its  ______________.	Every point on a circle is the same distance from its 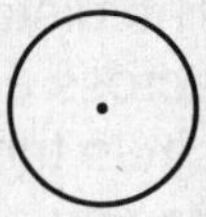 ______________.	The distance across a circle through its center is its 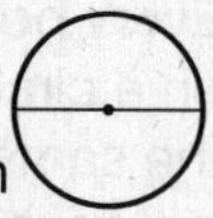 ______________.	The distance from a circle to its center is its 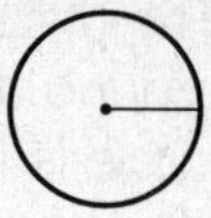 ______________.
Two or more circles with the same center are  ______________.	A segment between two points on a circle is a 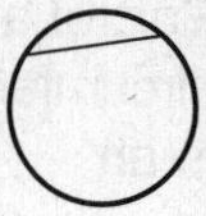 ______________.	Part of a circumference is an  ______________.	Part of a circle bounded by an arc and two radii is a 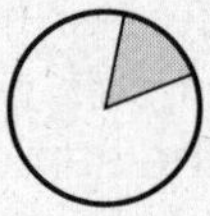 ______________.
Half a circle is a  ______________.	An angle whose vertex is the center of a circle is a  ______________.	An angle whose vertex is on the circle whose sides include chords is an 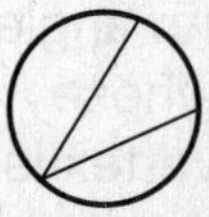 ______________.	A polygon whose vertices are on the circle and whose edges are within the circle is an 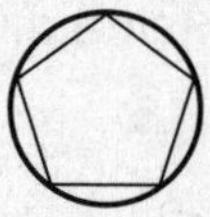 ______________.

Mental Math

a.	**b.**	**c.**	**d.**
e.	**f.**	**g.**	**h.**

Problem Solving

Understand
What information am I given?
What am I asked to find or do?

Plan
How can I use the information I am given?
Which strategy should I try?

Solve
Did I follow the plan?
Did I show my work?
Did I write the answer?

Check
Did I use the correct information?
Did I do what was asked?
Is my answer reasonable?

Name ______________________ Time ____________

Power Up E

Use with Lesson 23

Facts

Write the word or words to complete each definition.

The distance around a circle is its  ______________.	Every point on a circle is the same distance from its 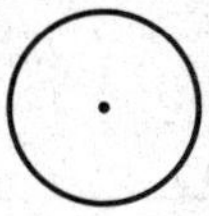 ______________.	The distance across a circle through its center is its 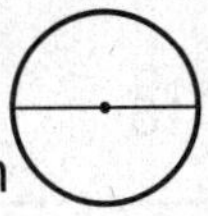 ______________.	The distance from a circle to its center is its 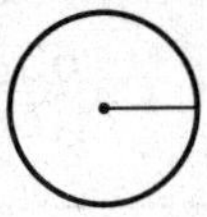 ______________.
Two or more circles with the same center are  ______________.	A segment between two points on a circle is a 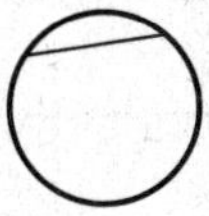 ______________.	Part of a circumference is an  ______________.	Part of a circle bounded by an arc and two radii is a 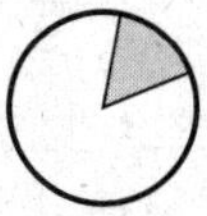 ______________.
Half a circle is a  ______________.	An angle whose vertex is the center of a circle is a  ______________.	An angle whose vertex is on the circle whose sides include chords is an  ______________.	A polygon whose vertices are on the circle and whose edges are within the circle is an 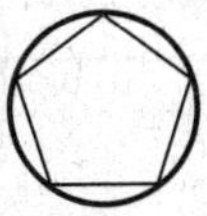 ______________.

Mental Math

a.	**b.**	**c.**	**d.**
e.	**f.**	**g.**	**h.**

Problem Solving

Understand

What information am I given?

What am I asked to find or do?

Plan

How can I use the information I am given?

Which strategy should I try?

Solve

Did I follow the plan?

Did I show my work?

Did I write the answer?

Check

Did I use the correct information?

Did I do what was asked?

Is my answer reasonable?

Name ______________________ Time ____________

Power Up D

Use with Lesson 24

Facts

Reduce each fraction to lowest terms.

$\frac{50}{100} =$	$\frac{4}{16} =$	$\frac{6}{8} =$	$\frac{8}{12} =$	$\frac{10}{100} =$
$\frac{8}{16} =$	$\frac{20}{100} =$	$\frac{3}{12} =$	$\frac{60}{100} =$	$\frac{9}{12} =$
$\frac{6}{9} =$	$\frac{90}{100} =$	$\frac{5}{10} =$	$\frac{12}{16} =$	$\frac{25}{100} =$
$\frac{4}{10} =$	$\frac{4}{6} =$	$\frac{75}{100} =$	$\frac{4}{12} =$	$\frac{6}{10} =$

Mental Math

a.	**b.**	**c.**	**d.**
e.	**f.**	**g.**	**h.**

Problem Solving

Understand

What information am I given?
What am I asked to find or do?

Plan

How can I use the information I am given?
Which strategy should I try?

Solve

Did I follow the plan?
Did I show my work?
Did I write the answer?

Check

Did I use the correct information?
Did I do what was asked?
Is my answer reasonable?

Name ____________________ Time __________

Power Up F

Use with Lesson 25

Facts

Name each figure illustrated.

1. 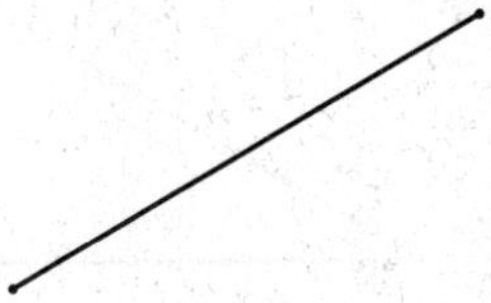 ______	2.  ______	3.  ______	4. 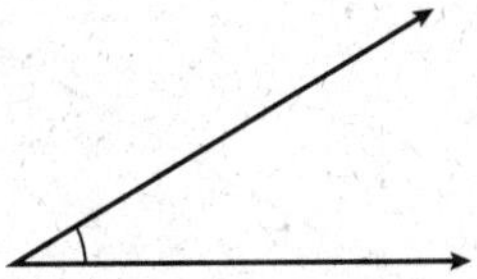 ______
5. 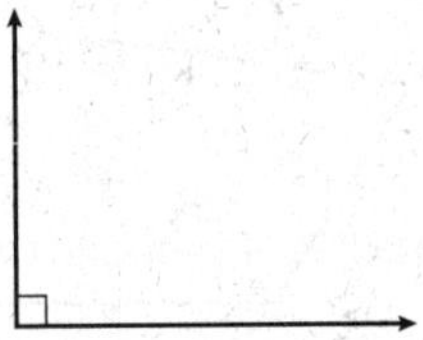 ______	6. 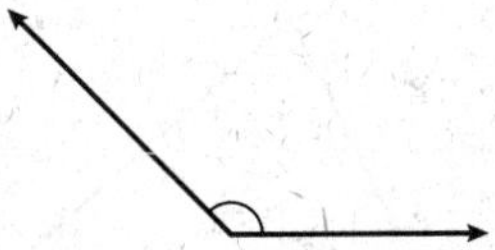 ______	7. 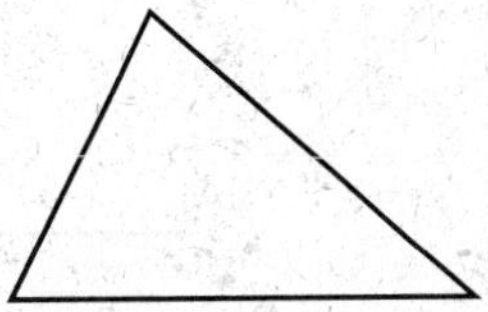 ______	8. 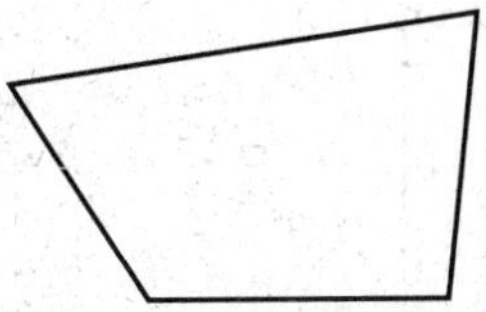 ______
9. 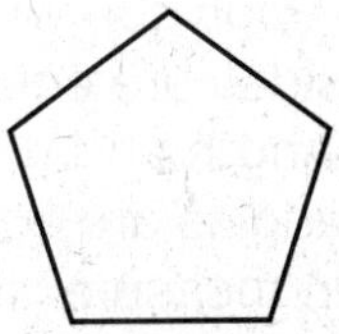 ______	10. 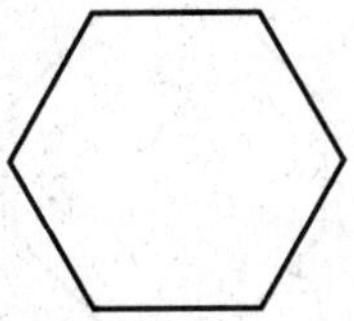 ______	11. 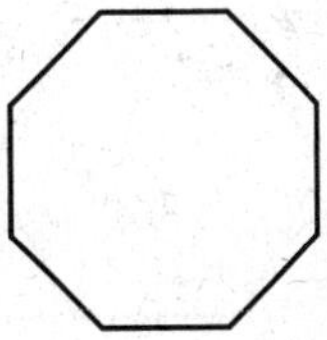 ______	12. A polygon whose sides are equal in length and whose angles are equal in measure is a ______.

Mental Math

a.	**b.**	**c.**	**d.**
e.	**f.**	**g.**	**h.**

Problem Solving

Understand

What information am I given?

What am I asked to find or do?

Plan

How can I use the information I am given?

Which strategy should I try?

Solve

Did I follow the plan?

Did I show my work?

Did I write the answer?

Check

Did I use the correct information?

Did I do what was asked?

Is my answer reasonable?

Name ______________________ Time ____________

Power Up F

Use with Lesson 26

Facts

Name each figure illustrated.

1. 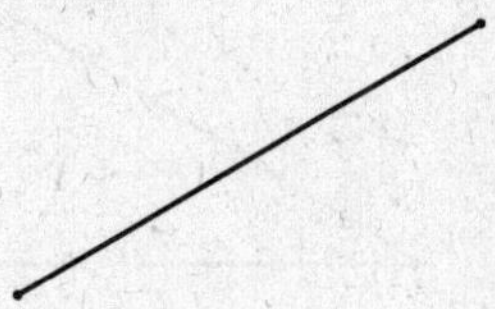 ______	2. ______	3. ______	4. 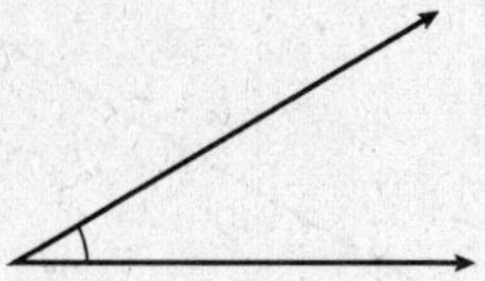 ______
5. 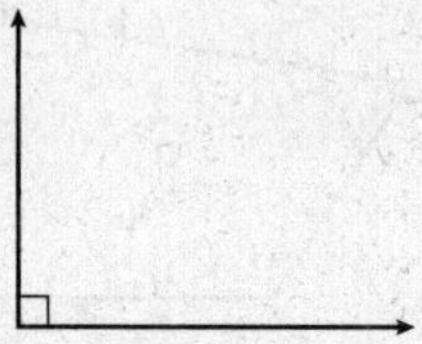 ______	6. 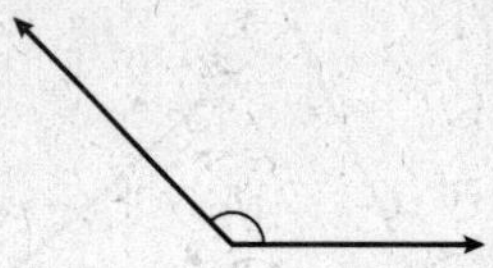 ______	7. 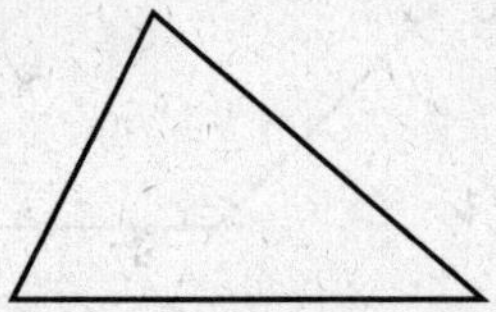 ______	8. 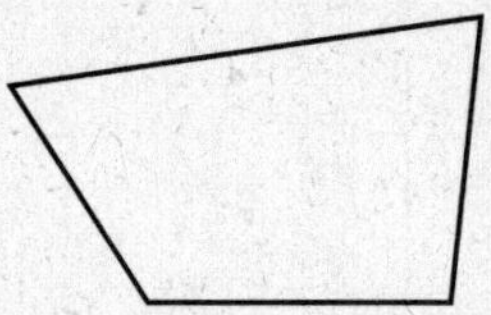 ______
9. 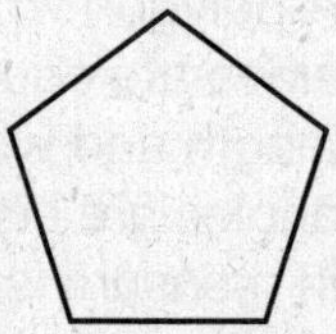 ______	10. 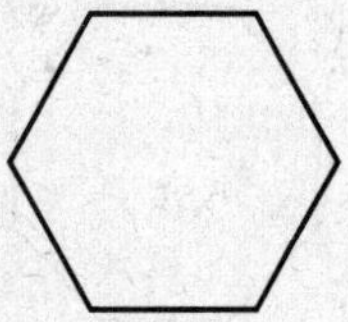 ______	11. 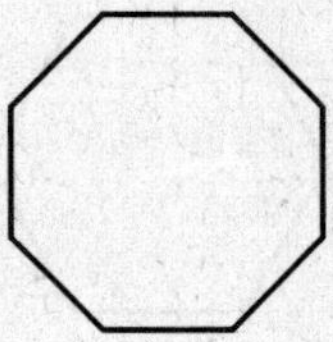 ______	12. A polygon whose sides are equal in length and whose angles are equal in measure is a ______.

Mental Math

a.	**b.**	**c.**	**d.**
e.	**f.**	**g.**	**h.**

Problem Solving

Understand
What information am I given?
What am I asked to find or do?

Plan
How can I use the information I am given?
Which strategy should I try?

Solve
Did I follow the plan?
Did I show my work?
Did I write the answer?

Check
Did I use the correct information?
Did I do what was asked?
Is my answer reasonable?

Name ____________________ Time __________

Power Up E

Use with Lesson 27

Facts

Write the word or words to complete each definition.

The distance around a circle is its 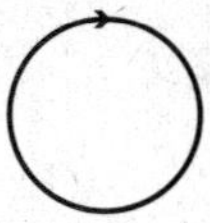 ______________.	Every point on a circle is the same distance from its 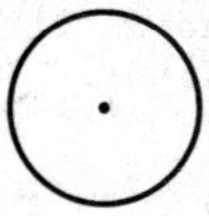 ______________.	The distance across a circle through its center is its 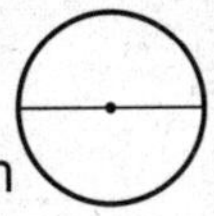 ______________.	The distance from a circle to its center is its 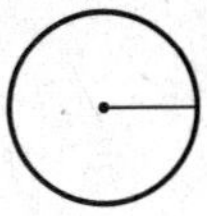 ______________.
Two or more circles with the same center are  ______________.	A segment between two points on a circle is a 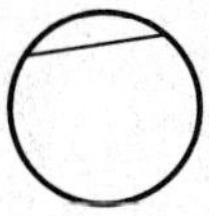 ______________.	Part of a circumference is an  ______________.	Part of a circle bounded by an arc and two radii is a 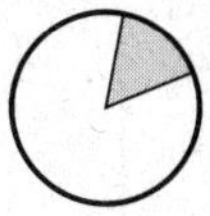 ______________.
Half a circle is a  ______________.	An angle whose vertex is the center of a circle is a  ______________.	An angle whose vertex is on the circle whose sides include chords is an 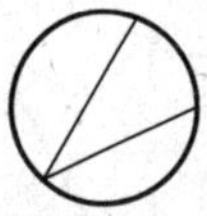 ______________.	A polygon whose vertices are on the circle and whose edges are within the circle is an 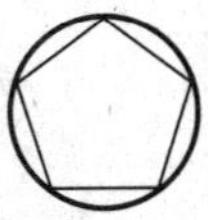 ______________.

Mental Math

a.	**b.**	**c.**	**d.**
e.	**f.**	**g.**	**h.**

Problem Solving

Understand

What information am I given?

What am I asked to find or do?

Plan

How can I use the information I am given?

Which strategy should I try?

Solve

Did I follow the plan?

Did I show my work?

Did I write the answer?

Check

Did I use the correct information?

Did I do what was asked?

Is my answer reasonable?

Name ______________________ Time __________

Power Up F

Use with Lesson 28

Facts

Name each figure illustrated.

1. 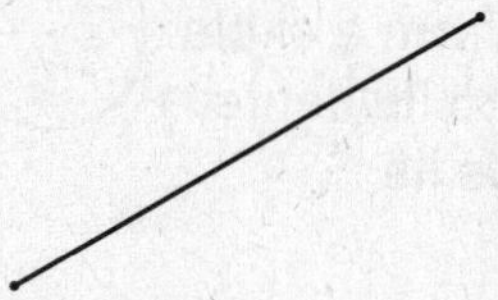 ____	2.  ____	3.  ____	4. 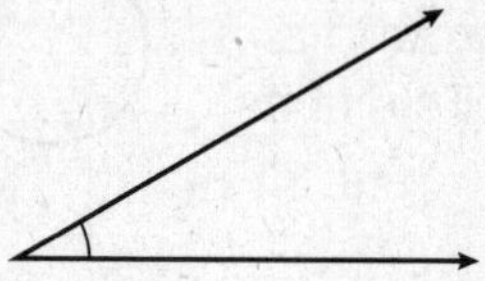 ____
5. 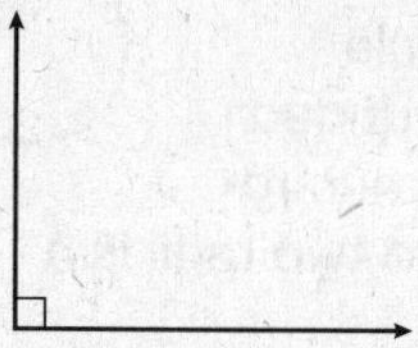 ____	6. 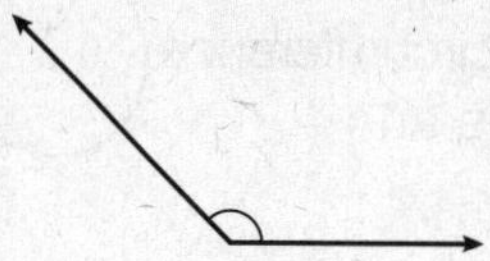 ____	7. 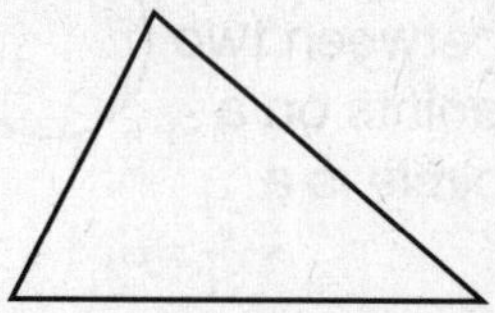 ____	8. 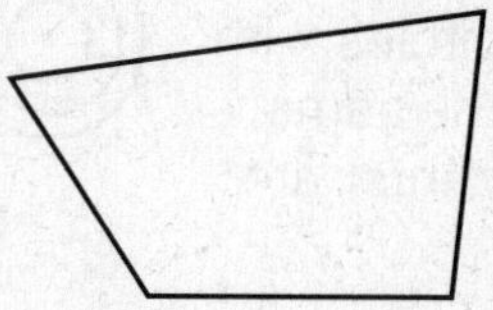 ____
9. 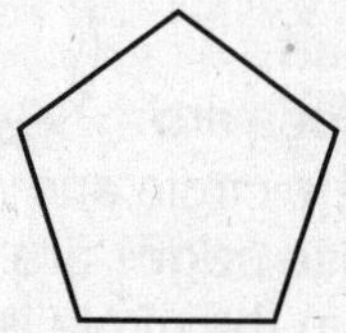 ____	10. 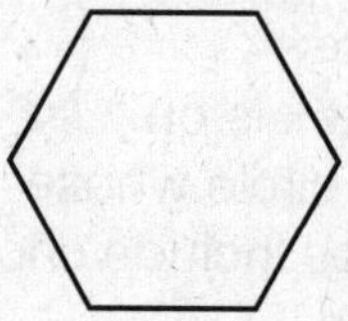 ____	11. 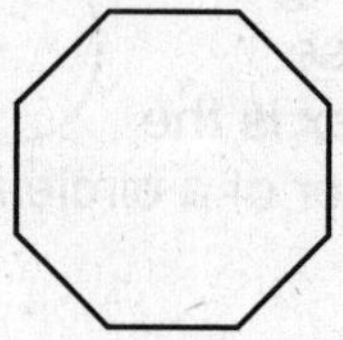 ____	12. A polygon whose sides are equal in length and whose angles are equal in measure is a ____.

Mental Math

a.	**b.**	**c.**	**d.**
e.	**f.**	**g.**	**h.**

Problem Solving

Understand

What information am I given?

What am I asked to find or do?

Plan

How can I use the information I am given?

Which strategy should I try?

Solve

Did I follow the plan?

Did I show my work?

Did I write the answer?

Check

Did I use the correct information?

Did I do what was asked?

Is my answer reasonable?

Name ______________________ Time __________

Power Up E

Use with Lesson 29

Facts

Write the word or words to complete each definition.

The distance around a circle is its 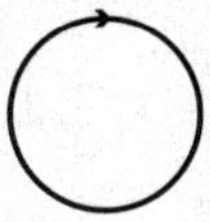 ________.	Every point on a circle is the same distance from its 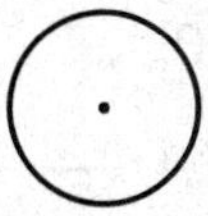 ________.	The distance across a circle through its center is its 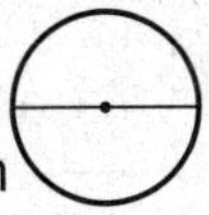 ________.	The distance from a circle to its center is its 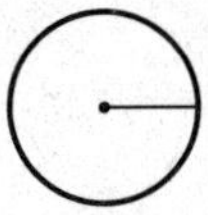 ________.
Two or more circles with the same center are  ________.	A segment between two points on a circle is a 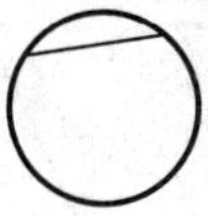 ________.	Part of a circumference is an  ________.	Part of a circle bounded by an arc and two radii is a 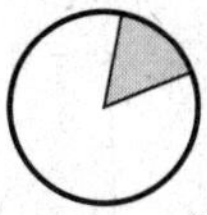 ________.
Half a circle is a  ________.	An angle whose vertex is the center of a circle is a  ________.	An angle whose vertex is on the circle whose sides include chords is an  ________.	A polygon whose vertices are on the circle and whose edges are within the circle is an 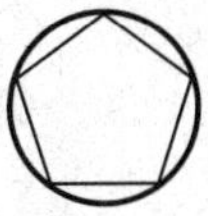 ________.

Mental Math

a.	b.	c.	d.
e.	f.	g.	h.

Problem Solving

Understand
What information am I given?
What am I asked to find or do?

Plan
How can I use the information I am given?
Which strategy should I try?

Solve
Did I follow the plan?
Did I show my work?
Did I write the answer?

Check
Did I use the correct information?
Did I do what was asked?
Is my answer reasonable?

Name ______________________ Time ____________

Power Up F

Use with Lesson 30

Facts

Name each figure illustrated.

1. 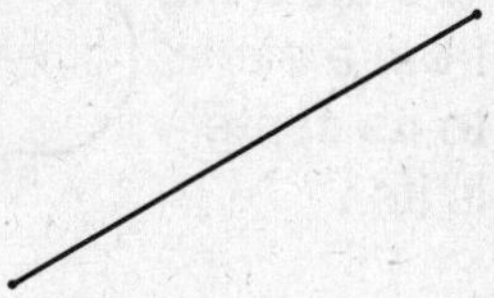 ______	2. ______	3. ______	4. 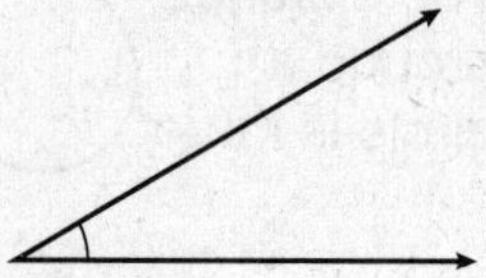 ______
5. 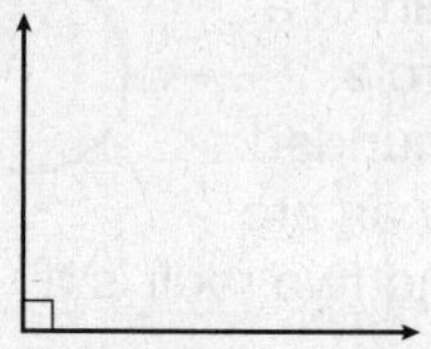 ______	6. 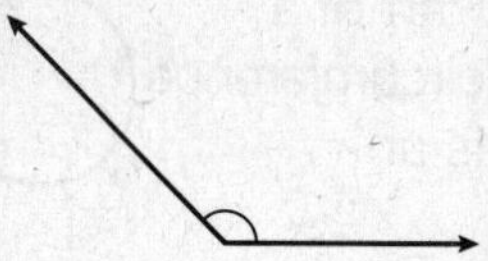 ______	7. 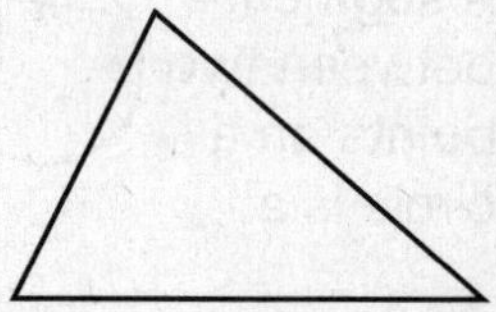 ______	8. 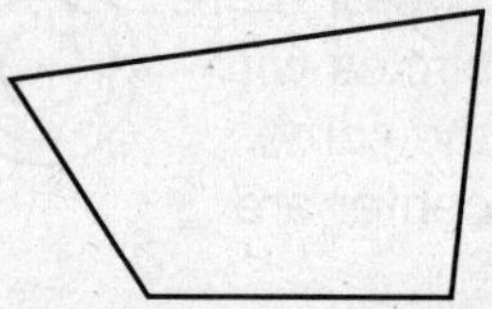 ______
9. 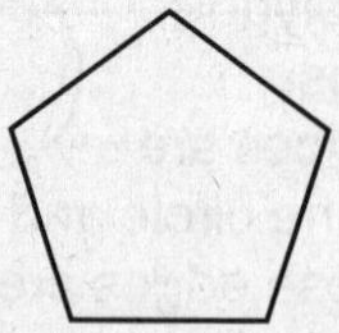 ______	10. 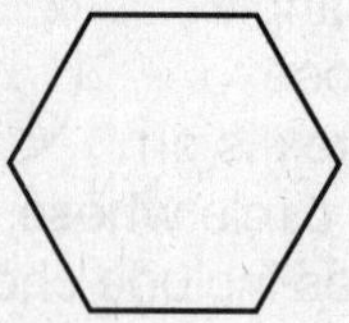 ______	11. 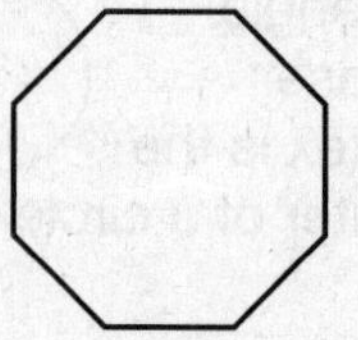 ______	12. A polygon whose sides are equal in length and whose angles are equal in measure is a ______.

Mental Math

a.	**b.**	**c.**	**d.**
e.	**f.**	**g.**	**h.**

Problem Solving

Understand
What information am I given?
What am I asked to find or do?

Plan
How can I use the information I am given?
Which strategy should I try?

Solve
Did I follow the plan?
Did I show my work?
Did I write the answer?

Check
Did I use the correct information?
Did I do what was asked?
Is my answer reasonable?

Name ____________________ Time __________

Power Up G

Use with Lesson 31

Facts

Simplify.

$\frac{2}{3} + \frac{2}{3} =$	$\frac{2}{3} - \frac{1}{3} =$	$\frac{2}{3} \times \frac{2}{3} =$	$\frac{2}{3} \div \frac{2}{3} =$
$\frac{3}{4} + \frac{1}{4} =$	$\frac{3}{4} - \frac{1}{4} =$	$\frac{3}{4} \times \frac{1}{4} =$	$\frac{3}{4} \div \frac{1}{4} =$
$\frac{2}{3} + \frac{1}{2} =$	$\frac{2}{3} - \frac{1}{2} =$	$\frac{2}{3} \times \frac{1}{2} =$	$\frac{2}{3} \div \frac{1}{2} =$
$\frac{3}{4} + \frac{2}{3} =$	$\frac{3}{4} - \frac{2}{3} =$	$\frac{3}{4} \times \frac{2}{3} =$	$\frac{3}{4} \div \frac{2}{3} =$

Mental Math

a.	**b.**	**c.**	**d.**
e.	**f.**	**g.**	**h.**

Problem Solving

Understand

What information am I given?

What am I asked to find or do?

Plan

How can I use the information I am given?

Which strategy should I try?

Solve

Did I follow the plan?

Did I show my work?

Did I write the answer?

Check

Did I use the correct information?

Did I do what was asked?

Is my answer reasonable?

Name ____________________ Time __________

Power Up G

Use with Lesson 32

Facts

Simplify.

$\frac{2}{3} + \frac{2}{3} =$	$\frac{2}{3} - \frac{1}{3} =$	$\frac{2}{3} \times \frac{2}{3} =$	$\frac{2}{3} \div \frac{2}{3} =$
$\frac{3}{4} + \frac{1}{4} =$	$\frac{3}{4} - \frac{1}{4} =$	$\frac{3}{4} \times \frac{1}{4} =$	$\frac{3}{4} \div \frac{1}{4} =$
$\frac{2}{3} + \frac{1}{2} =$	$\frac{2}{3} - \frac{1}{2} =$	$\frac{2}{3} \times \frac{1}{2} =$	$\frac{2}{3} \div \frac{1}{2} =$
$\frac{3}{4} + \frac{2}{3} =$	$\frac{3}{4} - \frac{2}{3} =$	$\frac{3}{4} \times \frac{2}{3} =$	$\frac{3}{4} \div \frac{2}{3} =$

Mental Math

a.	**b.**	**c.**	**d.**
e.	**f.**	**g.**	**h.**

Problem Solving

Understand

What information am I given?

What am I asked to find or do?

Plan

How can I use the information I am given?

Which strategy should I try?

Solve

Did I follow the plan?

Did I show my work?

Did I write the answer?

Check

Did I use the correct information?

Did I do what was asked?

Is my answer reasonable?

Name ______________________ Time ____________

Power Up F

Use with Lesson 33

Facts Name each figure illustrated.

1. 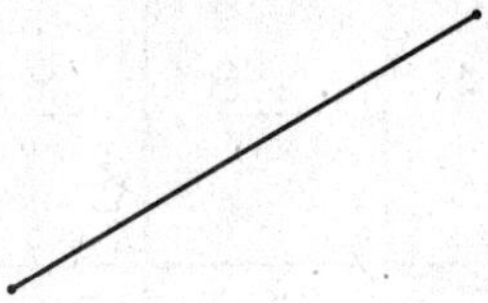 ____________	2. ____________	3. ____________	4. 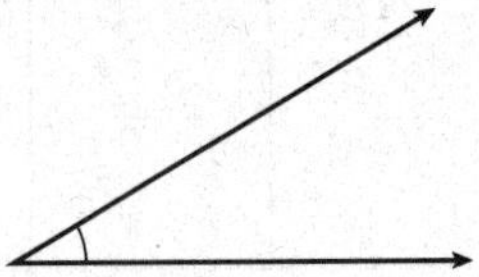 ____________
5. 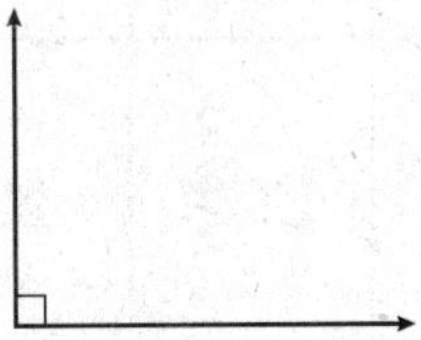 ____________	6. 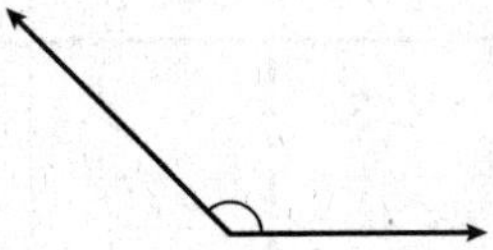 ____________	7. 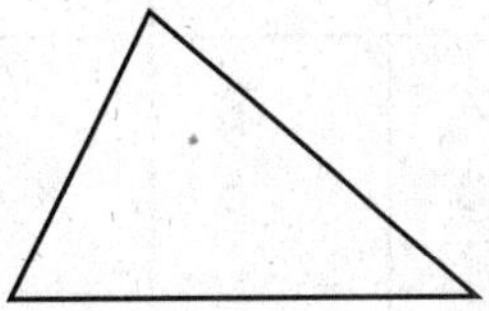 ____________	8. 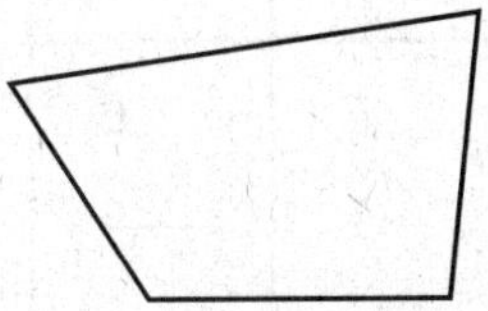 ____________
9. 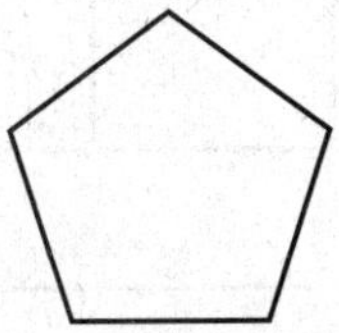 ____________	10. 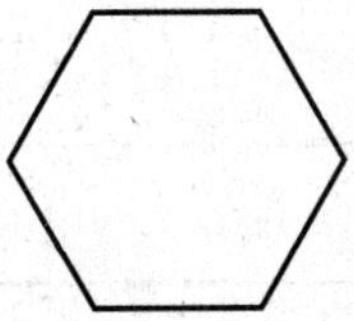 ____________	11. 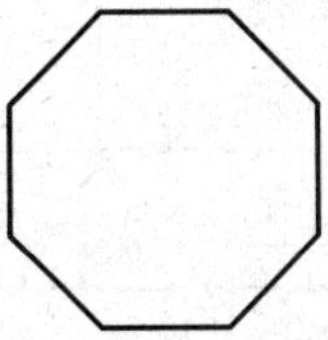 ____________	12. A polygon whose sides are equal in length and whose angles are equal in measure is a ____________.

Mental Math

a.	**b.**	**c.**	**d.**
e.	**f.**	**g.**	**h.**

Problem Solving

Understand
What information am I given?
What am I asked to find or do?

Plan
How can I use the information I am given?
Which strategy should I try?

Solve
Did I follow the plan?
Did I show my work?
Did I write the answer?

Check
Did I use the correct information?
Did I do what was asked?
Is my answer reasonable?

Name ____________________ Time ____________

Power Up A

Use with Lesson 34

Facts

Multiply.

9×8	8×2	10×10	6×3	4×2	5×5	9×9	6×4	9×6	7×3
9×3	6×5	0×0	7×6	8×8	7×4	5×3	9×7	2×2	8×6
7×7	6×2	4×3	8×5	4×4	3×2	$n \times 0$	8×4	6×6	9×2
8×3	5×4	$n \times 1$	7×2	9×5	8×7	3×3	9×4	5×2	7×5

Mental Math

a.	b.	c.	d.
e.	f.	g.	h.

Problem Solving

Understand

What information am I given?

What am I asked to find or do?

Plan

How can I use the information I am given?

Which strategy should I try?

Solve

Did I follow the plan?

Did I show my work?

Did I write the answer?

Check

Did I use the correct information?

Did I do what was asked?

Is my answer reasonable?

Name ______________________ Time ____________

Power Up H

Use with Lesson 35

Facts

Write the number that completes each equivalent measure.

1. 1 foot = ______ inches	15. 1 kilogram ≈ ______ pounds
2. 1 yard = ______ inches	16. 1 pint = ______ ounces
3. 1 yard = ______ feet	17. 1 pint = ______ cups
4. 1 mile = ______ feet	18. 1 quart = ______ pints
5. 1 centimeter = ______ millimeters	19. 1 gallon = ______ quarts
6. 1 meter = ______ millimeters	20. 1 liter = ______ milliliters
7. 1 meter = ______ centimeters	21–24. 1 milliliter of water has a volume of ______ and a mass of ______ .
8. 1 kilometer = ______ meters	One liter of water has a volume of ______ cm^3 and a mass of ______ kg.
9. 1 inch = ______ centimeters	
10. 1 pound = ______ ounces	25–26. Water freezes at ____ °F and ____ °C.
11. 1 ton = ______ pounds	27–28. Water boils at ____ °F and ____ °C.
12. 1 gram = ______ milligrams	29–30. Normal body temperature is ____ °F and ______ °C.
13. 1 kilogram = ______ grams	
14. 1 metric ton = ______ kilograms	

Mental Math

a.	**b.**	**c.**	**d.**
e.	**f.**	**g.**	**h.**

Problem Solving

Understand
What information am I given?
What am I asked to find or do?

Plan
How can I use the information I am given?
Which strategy should I try?

Solve
Did I follow the plan?
Did I show my work?
Did I write the answer?

Check
Did I use the correct information?
Did I do what was asked?
Is my answer reasonable?

Name ______________________ Time __________

Power Up G

Use with Lesson 36

Facts Simplify.

$\frac{2}{3} + \frac{2}{3} =$	$\frac{2}{3} - \frac{1}{3} =$	$\frac{2}{3} \times \frac{2}{3} =$	$\frac{2}{3} \div \frac{2}{3} =$
$\frac{3}{4} + \frac{1}{4} =$	$\frac{3}{4} - \frac{1}{4} =$	$\frac{3}{4} \times \frac{1}{4} =$	$\frac{3}{4} \div \frac{1}{4} =$
$\frac{2}{3} + \frac{1}{2} =$	$\frac{2}{3} - \frac{1}{2} =$	$\frac{2}{3} \times \frac{1}{2} =$	$\frac{2}{3} \div \frac{1}{2} =$
$\frac{3}{4} + \frac{2}{3} =$	$\frac{3}{4} - \frac{2}{3} =$	$\frac{3}{4} \times \frac{2}{3} =$	$\frac{3}{4} \div \frac{2}{3} =$

Mental Math

a.	**b.**	**c.**	**d.**
e.	**f.**	**g.**	**h.**

Problem Solving

Understand

What information am I given?

What am I asked to find or do?

Plan

How can I use the information I am given?

Which strategy should I try?

Solve

Did I follow the plan?

Did I show my work?

Did I write the answer?

Check

Did I use the correct information?

Did I do what was asked?

Is my answer reasonable?

Name ______________________ Time __________

Power Up H

Use with Lesson 37

Facts Write the number that completes each equivalent measure.

1. 1 foot = ______ inches	15. 1 kilogram ≈ ______ pounds
2. 1 yard = ______ inches	16. 1 pint = ______ ounces
3. 1 yard = ______ feet	17. 1 pint = ______ cups
4. 1 mile = ______ feet	18. 1 quart = ______ pints
5. 1 centimeter = ______ millimeters	19. 1 gallon = ______ quarts
6. 1 meter = ______ millimeters	20. 1 liter = ______ milliliters
7. 1 meter = ______ centimeters	21–24. 1 milliliter of water has a volume of ______ and a mass of ______ .
8. 1 kilometer = ______ meters	One liter of water has a volume of ______ cm^3 and a mass of ______ kg.
9. 1 inch = ______ centimeters	
10. 1 pound = ______ ounces	25–26. Water freezes at ____ °F and ____ °C.
11. 1 ton = ______ pounds	27–28. Water boils at ____ °F and ____ °C.
12. 1 gram = ______ milligrams	29–30. Normal body temperature is ____ °F and ______ °C.
13. 1 kilogram = ______ grams	
14. 1 metric ton = ______ kilograms	

Mental Math

a.	b.	c.	d.
e.	f.	g.	h.

Problem Solving

Understand
What information am I given?
What am I asked to find or do?

Plan
How can I use the information I am given?
Which strategy should I try?

Solve
Did I follow the plan?
Did I show my work?
Did I write the answer?

Check
Did I use the correct information?
Did I do what was asked?
Is my answer reasonable?

Name ______________________ Time ____________

Power Up G

Use with Lesson 38

Facts

Simplify.

$\frac{2}{3} + \frac{2}{3} =$	$\frac{2}{3} - \frac{1}{3} =$	$\frac{2}{3} \times \frac{2}{3} =$	$\frac{2}{3} \div \frac{2}{3} =$
$\frac{3}{4} + \frac{1}{4} =$	$\frac{3}{4} - \frac{1}{4} =$	$\frac{3}{4} \times \frac{1}{4} =$	$\frac{3}{4} \div \frac{1}{4} =$
$\frac{2}{3} + \frac{1}{2} =$	$\frac{2}{3} - \frac{1}{2} =$	$\frac{2}{3} \times \frac{1}{2} =$	$\frac{2}{3} \div \frac{1}{2} =$
$\frac{3}{4} + \frac{2}{3} =$	$\frac{3}{4} - \frac{2}{3} =$	$\frac{3}{4} \times \frac{2}{3} =$	$\frac{3}{4} \div \frac{2}{3} =$

Mental Math

a.	**b.**	**c.**	**d.**
e.	**f.**	**g.**	**h.**

Problem Solving

Understand

What information am I given?

What am I asked to find or do?

Plan

How can I use the information I am given?

Which strategy should I try?

Solve

Did I follow the plan?

Did I show my work?

Did I write the answer?

Check

Did I use the correct information?

Did I do what was asked?

Is my answer reasonable?

Name ____________________ Time __________

Power Up H

Use with Lesson 39

Facts

Write the number that completes each equivalent measure.

1. 1 foot = ______ inches	15. 1 kilogram ≈ ______ pounds
2. 1 yard = ______ inches	16. 1 pint = ______ ounces
3. 1 yard = ______ feet	17. 1 pint = ______ cups
4. 1 mile = ______ feet	18. 1 quart = ______ pints
5. 1 centimeter = ______ millimeters	19. 1 gallon = ______ quarts
6. 1 meter = ______ millimeters	20. 1 liter = ______ milliliters
7. 1 meter = ______ centimeters	21–24. 1 milliliter of water has a volume of ______ and a mass of ______ .
8. 1 kilometer = ______ meters	One liter of water has a volume of ______ cm^3 and a mass of ______ kg.
9. 1 inch = ______ centimeters	
10. 1 pound = ______ ounces	25–26. Water freezes at ____ °F and ____ °C.
11. 1 ton = ______ pounds	27–28. Water boils at ____ °F and ____ °C.
12. 1 gram = ______ milligrams	29–30. Normal body temperature is ____ °F and ______ °C.
13. 1 kilogram = ______ grams	
14. 1 metric ton = ______ kilograms	

Mental Math

a.	b.	c.	d.
e.	f.	g.	h.

Problem Solving

Understand
What information am I given?
What am I asked to find or do?

Plan
How can I use the information I am given?
Which strategy should I try?

Solve
Did I follow the plan?
Did I show my work?
Did I write the answer?

Check
Did I use the correct information?
Did I do what was asked?
Is my answer reasonable?

Name ______________________ Time __________

Power Up G

Use with Lesson 40

Facts

Simplify.

$\frac{2}{3} + \frac{2}{3} =$	$\frac{2}{3} - \frac{1}{3} =$	$\frac{2}{3} \times \frac{2}{3} =$	$\frac{2}{3} \div \frac{2}{3} =$
$\frac{3}{4} + \frac{1}{4} =$	$\frac{3}{4} - \frac{1}{4} =$	$\frac{3}{4} \times \frac{1}{4} =$	$\frac{3}{4} \div \frac{1}{4} =$
$\frac{2}{3} + \frac{1}{2} =$	$\frac{2}{3} - \frac{1}{2} =$	$\frac{2}{3} \times \frac{1}{2} =$	$\frac{2}{3} \div \frac{1}{2} =$
$\frac{3}{4} + \frac{2}{3} =$	$\frac{3}{4} - \frac{2}{3} =$	$\frac{3}{4} \times \frac{2}{3} =$	$\frac{3}{4} \div \frac{2}{3} =$

Mental Math

a.	**b.**	**c.**	**d.**
e.	**f.**	**g.**	**h.**

Problem Solving

Understand

What information am I given?

What am I asked to find or do?

Plan

How can I use the information I am given?

Which strategy should I try?

Solve

Did I follow the plan?

Did I show my work?

Did I write the answer?

Check

Did I use the correct information?

Did I do what was asked?

Is my answer reasonable?

Name ____________________ Time __________

Power Up A

Use with Lesson 41

Facts Multiply.

9×8	8×2	10×10	6×3	4×2	5×5	9×9	6×4	9×6	7×3
9×3	6×5	0×0	7×6	8×8	7×4	5×3	9×7	2×2	8×6
7×7	6×2	4×3	8×5	4×4	3×2	$n \times 0$	8×4	6×6	9×2
8×3	5×4	$n \times 1$	7×2	9×5	8×7	3×3	9×4	5×2	7×5

Mental Math

a.	**b.**	**c.**	**d.**
e.	**f.**	**g.**	**h.**

Problem Solving

Understand
What information am I given?
What am I asked to find or do?

Plan
How can I use the information I am given?
Which strategy should I try?

Solve
Did I follow the plan?
Did I show my work?
Did I write the answer?

Check
Did I use the correct information?
Did I do what was asked?
Is my answer reasonable?

Name ______________________ Time __________

Power Up H

Use with Lesson 42

Facts

Write the number that completes each equivalent measure.

1. 1 foot = ______ inches	15. 1 kilogram ≈ ______ pounds
2. 1 yard = ______ inches	16. 1 pint = ______ ounces
3. 1 yard = ______ feet	17. 1 pint = ______ cups
4. 1 mile = ______ feet	18. 1 quart = ______ pints
5. 1 centimeter = ______ millimeters	19. 1 gallon = ______ quarts
6. 1 meter = ______ millimeters	20. 1 liter = ______ milliliters
7. 1 meter = ______ centimeters	21–24. 1 milliliter of water has a volume of ______ and a mass of ______ .
8. 1 kilometer = ______ meters	One liter of water has a volume of ______ cm^3 and a mass of ______ kg.
9. 1 inch = ______ centimeters	
10. 1 pound = ______ ounces	25–26. Water freezes at ____ °F and ____ °C.
11. 1 ton = ______ pounds	27–28. Water boils at ____ °F and ____ °C.
12. 1 gram = ______ milligrams	29–30. Normal body temperature is ____ °F and ______ °C.
13. 1 kilogram = ______ grams	
14. 1 metric ton = ______ kilograms	

Mental Math

a.	b.	c.	d.
e.	f.	g.	h.

Problem Solving

Understand
What information am I given?
What am I asked to find or do?

Plan
How can I use the information I am given?
Which strategy should I try?

Solve
Did I follow the plan?
Did I show my work?
Did I write the answer?

Check
Did I use the correct information?
Did I do what was asked?
Is my answer reasonable?

Name ______________________ Time ____________

Power Up I

Use with Lesson 43

Facts

Find the number that completes each proportion.

$\frac{3}{4} = \frac{a}{12}$	$\frac{3}{4} = \frac{12}{b}$	$\frac{c}{5} = \frac{12}{20}$	$\frac{2}{d} = \frac{12}{24}$	$\frac{8}{12} = \frac{4}{e}$
$\frac{f}{10} = \frac{10}{5}$	$\frac{5}{g} = \frac{25}{100}$	$\frac{10}{100} = \frac{5}{h}$	$\frac{8}{4} = \frac{j}{16}$	$\frac{24}{k} = \frac{8}{6}$
$\frac{9}{12} = \frac{36}{m}$	$\frac{50}{100} = \frac{w}{30}$	$\frac{3}{9} = \frac{5}{p}$	$\frac{q}{60} = \frac{15}{20}$	$\frac{2}{5} = \frac{r}{100}$

Mental Math

a.	**b.**	**c.**	**d.**
e.	**f.**	**g.**	**h.**

Problem Solving

Understand

What information am I given?

What am I asked to find or do?

Plan

How can I use the information I am given?

Which strategy should I try?

Solve

Did I follow the plan?

Did I show my work?

Did I write the answer?

Check

Did I use the correct information?

Did I do what was asked?

Is my answer reasonable?

Name ______________ Time ______

Power Up H

Use with Lesson 44

Facts Write the number that completes each equivalent measure.

1. 1 foot = ______ inches	15. 1 kilogram ≈ ______ pounds
2. 1 yard = ______ inches	16. 1 pint = ______ ounces
3. 1 yard = ______ feet	17. 1 pint = ______ cups
4. 1 mile = ______ feet	18. 1 quart = ______ pints
5. 1 centimeter = ______ millimeters	19. 1 gallon = ______ quarts
6. 1 meter = ______ millimeters	20. 1 liter = ______ milliliters
7. 1 meter = ______ centimeters	21–24. 1 milliliter of water has a volume of ______ and a mass of ______ . One liter of water has a volume of ______ cm^3 and a mass of ______ kg.
8. 1 kilometer = ______ meters	
9. 1 inch = ______ centimeters	
10. 1 pound = ______ ounces	25–26. Water freezes at ____ °F and ____ °C.
11. 1 ton = ______ pounds	27–28. Water boils at ____ °F and ____ °C.
12. 1 gram = ______ milligrams	29–30. Normal body temperature is ____ °F and ______ °C.
13. 1 kilogram = ______ grams	
14. 1 metric ton = ______ kilograms	

Mental Math

a.	b.	c.	d.
e.	f.	g.	h.

Problem Solving

Understand

What information am I given?

What am I asked to find or do?

Plan

How can I use the information I am given?

Which strategy should I try?

Solve

Did I follow the plan?

Did I show my work?

Did I write the answer?

Check

Did I use the correct information?

Did I do what was asked?

Is my answer reasonable?

Facts

Find the number that completes each proportion.

$\frac{3}{4} = \frac{a}{12}$	$\frac{3}{4} = \frac{12}{b}$	$\frac{c}{5} = \frac{12}{20}$	$\frac{2}{d} = \frac{12}{24}$	$\frac{8}{12} = \frac{4}{e}$
$\frac{f}{10} = \frac{10}{5}$	$\frac{5}{g} = \frac{25}{100}$	$\frac{10}{100} = \frac{5}{h}$	$\frac{8}{4} = \frac{j}{16}$	$\frac{24}{k} = \frac{8}{6}$
$\frac{9}{12} = \frac{36}{m}$	$\frac{50}{100} = \frac{w}{30}$	$\frac{3}{9} = \frac{5}{p}$	$\frac{q}{60} = \frac{15}{20}$	$\frac{2}{5} = \frac{r}{100}$

Mental Math

a.	**b.**	**c.**	**d.**
e.	**f.**	**g.**	**h.**

Problem Solving

Understand
What information am I given?
What am I asked to find or do?

Plan
How can I use the information I am given?
Which strategy should I try?

Solve
Did I follow the plan?
Did I show my work?
Did I write the answer?

Check
Did I use the correct information?
Did I do what was asked?
Is my answer reasonable?

Name ______________________ Time ____________

Power Up I

Use with Lesson 46

Facts Find the number that completes each proportion.

$\frac{3}{4} = \frac{a}{12}$	$\frac{3}{4} = \frac{12}{b}$	$\frac{c}{5} = \frac{12}{20}$	$\frac{2}{d} = \frac{12}{24}$	$\frac{8}{12} = \frac{4}{e}$
$\frac{f}{10} = \frac{10}{5}$	$\frac{5}{g} = \frac{25}{100}$	$\frac{10}{100} = \frac{5}{h}$	$\frac{8}{4} = \frac{j}{16}$	$\frac{24}{k} = \frac{8}{6}$
$\frac{9}{12} = \frac{36}{m}$	$\frac{50}{100} = \frac{w}{30}$	$\frac{3}{9} = \frac{5}{p}$	$\frac{q}{60} = \frac{15}{20}$	$\frac{2}{5} = \frac{r}{100}$

Mental Math

a.	**b.**	**c.**	**d.**
e.	**f.**	**g.**	**h.**

Problem Solving

Understand
What information am I given?
What am I asked to find or do?

Plan
How can I use the information I am given?
Which strategy should I try?

Solve
Did I follow the plan?
Did I show my work?
Did I write the answer?

Check
Did I use the correct information?
Did I do what was asked?
Is my answer reasonable?

Name ______________________ Time ____________

Power Up I

Use with Lesson 47

Facts

Find the number that completes each proportion.

$\frac{3}{4} = \frac{a}{12}$	$\frac{3}{4} = \frac{12}{b}$	$\frac{c}{5} = \frac{12}{20}$	$\frac{2}{d} = \frac{12}{24}$	$\frac{8}{12} = \frac{4}{e}$
$\frac{f}{10} = \frac{10}{5}$	$\frac{5}{g} = \frac{25}{100}$	$\frac{10}{100} = \frac{5}{h}$	$\frac{8}{4} = \frac{j}{16}$	$\frac{24}{k} = \frac{8}{6}$
$\frac{9}{12} = \frac{36}{m}$	$\frac{50}{100} = \frac{w}{30}$	$\frac{3}{9} = \frac{5}{p}$	$\frac{q}{60} = \frac{15}{20}$	$\frac{2}{5} = \frac{r}{100}$

Mental Math

a.	**b.**	**c.**	**d.**
e.	**f.**	**g.**	**h.**

Problem Solving

Understand

What information am I given?

What am I asked to find or do?

Plan

How can I use the information I am given?

Which strategy should I try?

Solve

Did I follow the plan?

Did I show my work?

Did I write the answer?

Check

Did I use the correct information?

Did I do what was asked?

Is my answer reasonable?

Name ______________________ Time ____________

Power Up J

Use with Lesson 48

Facts Simplify.

0.8 + 0.4 =	0.8 – 0.4 =	0.8 × 0.4 =	0.8 ÷ 0.4 =
1.2 + 0.4 =	1.2 – 0.4 =	1.2 × 0.4 =	1.2 ÷ 0.4 =
6 + 0.3 =	6 – 0.3 =	6 × 0.3 =	6 ÷ 0.3 =
1.2 + 4 =	0.01 – 0.01 =	0.3 × 0.3 =	0.12 ÷ 4 =

Mental Math

a.	**b.**	**c.**	**d.**
e.	**f.**	**g.**	**h.**

Problem Solving

Understand

What information am I given?

What am I asked to find or do?

Plan

How can I use the information I am given?

Which strategy should I try?

Solve

Did I follow the plan?

Did I show my work?

Did I write the answer?

Check

Did I use the correct information?

Did I do what was asked?

Is my answer reasonable?

Name ______________________ Time __________

Power Up J

Use with Lesson 49

Facts

Simplify.

0.8 + 0.4 =	0.8 − 0.4 =	0.8 × 0.4 =	0.8 ÷ 0.4 =
1.2 + 0.4 =	1.2 − 0.4 =	1.2 × 0.4 =	1.2 ÷ 0.4 =
6 + 0.3 =	6 − 0.3 =	6 × 0.3 =	6 ÷ 0.3 =
1.2 + 4 =	0.01 − 0.01 =	0.3 × 0.3 =	0.12 ÷ 4 =

Mental Math

a.	**b.**	**c.**	**d.**
e.	**f.**	**g.**	**h.**

Problem Solving

Understand
What information am I given?
What am I asked to find or do?

Plan
How can I use the information I am given?
Which strategy should I try?

Solve
Did I follow the plan?
Did I show my work?
Did I write the answer?

Check
Did I use the correct information?
Did I do what was asked?
Is my answer reasonable?

Name ______________________ Time __________

Power Up I

Use with Lesson 50

Facts Find the number that completes each proportion.

$\frac{3}{4} = \frac{a}{12}$	$\frac{3}{4} = \frac{12}{b}$	$\frac{c}{5} = \frac{12}{20}$	$\frac{2}{d} = \frac{12}{24}$	$\frac{8}{12} = \frac{4}{e}$
$\frac{f}{10} = \frac{10}{5}$	$\frac{5}{g} = \frac{25}{100}$	$\frac{10}{100} = \frac{5}{h}$	$\frac{8}{4} = \frac{j}{16}$	$\frac{24}{k} = \frac{8}{6}$
$\frac{9}{12} = \frac{36}{m}$	$\frac{50}{100} = \frac{w}{30}$	$\frac{3}{9} = \frac{5}{p}$	$\frac{q}{60} = \frac{15}{20}$	$\frac{2}{5} = \frac{r}{100}$

Mental Math

a.	**b.**	**c.**	**d.**
e.	**f.**	**g.**	**h.**

Problem Solving

Understand
What information am I given?
What am I asked to find or do?

Plan
How can I use the information I am given?
Which strategy should I try?

Solve
Did I follow the plan?
Did I show my work?
Did I write the answer?

Check
Did I use the correct information?
Did I do what was asked?
Is my answer reasonable?

Name ______________________ Time ____________

Power Up J

Use with Lesson 51

Facts

Simplify.

$0.8 + 0.4 =$	$0.8 - 0.4 =$	$0.8 \times 0.4 =$	$0.8 \div 0.4 =$
$1.2 + 0.4 =$	$1.2 - 0.4 =$	$1.2 \times 0.4 =$	$1.2 \div 0.4 =$
$6 + 0.3 =$	$6 - 0.3 =$	$6 \times 0.3 =$	$6 \div 0.3 =$
$1.2 + 4 =$	$0.01 - 0.01 =$	$0.3 \times 0.3 =$	$0.12 \div 4 =$

Mental Math

a.	**b.**	**c.**	**d.**
e.	**f.**	**g.**	**h.**

Problem Solving

Understand

What information am I given?

What am I asked to find or do?

Plan

How can I use the information I am given?

Which strategy should I try?

Solve

Did I follow the plan?

Did I show my work?

Did I write the answer?

Check

Did I use the correct information?

Did I do what was asked?

Is my answer reasonable?

Name ______________________ Time __________

Power Up K

Use with Lesson 52

Facts Simplify each power or root.

$\sqrt{100}$ =	$\sqrt{16}$ =	$\sqrt{81}$ =	$\sqrt{4}$ =	$\sqrt{144}$ =
$\sqrt{64}$ =	$\sqrt{49}$ =	$\sqrt{25}$ =	$\sqrt{9}$ =	$\sqrt{36}$ =
8^2 =	5^2 =	3^2 =	12^2 =	10^2 =
7^2 =	2^3 =	3^3 =	10^3 =	5^3 =

Mental Math

a.	**b.**	**c.**	**d.**
e.	**f.**	**g.**	**h.**

Problem Solving

Understand

What information am I given?

What am I asked to find or do?

Plan

How can I use the information I am given?

Which strategy should I try?

Solve

Did I follow the plan?

Did I show my work?

Did I write the answer?

Check

Did I use the correct information?

Did I do what was asked?

Is my answer reasonable?

Name ______________________ Time __________

Power Up A

Use with Lesson 53

Facts Multiply.

9 × 8	8 × 2	10 × 10	6 × 3	4 × 2	5 × 5	9 × 9	6 × 4	9 × 6	7 × 3
9 × 3	6 × 5	0 × 0	7 × 6	8 × 8	7 × 4	5 × 3	9 × 7	2 × 2	8 × 6
7 × 7	6 × 2	4 × 3	8 × 5	4 × 4	3 × 2	n × 0	8 × 4	6 × 6	9 × 2
8 × 3	5 × 4	n × 1	7 × 2	9 × 5	8 × 7	3 × 3	9 × 4	5 × 2	7 × 5

Mental Math

a.	**b.**	**c.**	**d.**
e.	**f.**	**g.**	**h.**

Problem Solving

Understand
What information am I given?
What am I asked to find or do?

Plan
How can I use the information I am given?
Which strategy should I try?

Solve
Did I follow the plan?
Did I show my work?
Did I write the answer?

Check
Did I use the correct information?
Did I do what was asked?
Is my answer reasonable?

Name ____________________ Time ____________

Power Up L

Use with Lesson 54

Facts Write the equivalent decimal and percent for each fraction.

Fraction	Decimal	Percent	Fraction	Decimal	Percent
$\frac{1}{2}$			$\frac{1}{8}$		
$\frac{1}{3}$			$\frac{1}{10}$		
$\frac{2}{3}$			$\frac{3}{10}$		
$\frac{1}{4}$			$\frac{9}{10}$		
$\frac{3}{4}$			$\frac{1}{100}$		
$\frac{1}{5}$			$1\frac{1}{2}$		

Mental Math

a.	**b.**	**c.**	**d.**
e.	**f.**	**g.**	**h.**

Problem Solving

Understand
What information am I given?
What am I asked to find or do?

Plan
How can I use the information I am given?
Which strategy should I try?

Solve
Did I follow the plan?
Did I show my work?
Did I write the answer?

Check
Did I use the correct information?
Did I do what was asked?
Is my answer reasonable?

Name ____________________ Time __________

Power Up J

Use with Lesson 55

Facts

Simplify.

0.8 + 0.4 =	0.8 – 0.4 =	0.8 × 0.4 =	0.8 ÷ 0.4 =
1.2 + 0.4 =	1.2 – 0.4 =	1.2 × 0.4 =	1.2 ÷ 0.4 =
6 + 0.3 =	6 – 0.3 =	6 × 0.3 =	6 ÷ 0.3 =
1.2 + 4 =	0.01 – 0.01 =	0.3 × 0.3 =	0.12 ÷ 4 =

Mental Math

a.	**b.**	**c.**	**d.**
e.	**f.**	**g.**	**h.**

Problem Solving

Understand

What information am I given?

What am I asked to find or do?

Plan

How can I use the information I am given?

Which strategy should I try?

Solve

Did I follow the plan?

Did I show my work?

Did I write the answer?

Check

Did I use the correct information?

Did I do what was asked?

Is my answer reasonable?

Name ____________________ Time __________

Power Up L

Use with Lesson 56

Facts Write the equivalent decimal and percent for each fraction.

Fraction	Decimal	Percent	Fraction	Decimal	Percent
$\frac{1}{2}$			$\frac{1}{8}$		
$\frac{1}{3}$			$\frac{1}{10}$		
$\frac{2}{3}$			$\frac{3}{10}$		
$\frac{1}{4}$			$\frac{9}{10}$		
$\frac{3}{4}$			$\frac{1}{100}$		
$\frac{1}{5}$			$1\frac{1}{2}$		

Mental Math

a.	**b.**	**c.**	**d.**
e.	**f.**	**g.**	**h.**

Problem Solving

Understand

What information am I given?

What am I asked to find or do?

Plan

How can I use the information I am given?

Which strategy should I try?

Solve

Did I follow the plan?

Did I show my work?

Did I write the answer?

Check

Did I use the correct information?

Did I do what was asked?

Is my answer reasonable?

Name ____________________ Time __________

Power Up A

Use with Lesson 57

Facts

Multiply.

9×8	8×2	10×10	6×3	4×2	5×5	9×9	6×4	9×6	7×3
9×3	6×5	0×0	7×6	8×8	7×4	5×3	9×7	2×2	8×6
7×7	6×2	4×3	8×5	4×4	3×2	$n \times 0$	8×4	6×6	9×2
8×3	5×4	$n \times 1$	7×2	9×5	8×7	3×3	9×4	5×2	7×5

Mental Math

a.	**b.**	**c.**	**d.**
e.	**f.**	**g.**	**h.**

Problem Solving

Understand
What information am I given?
What am I asked to find or do?

Plan
How can I use the information I am given?
Which strategy should I try?

Solve
Did I follow the plan?
Did I show my work?
Did I write the answer?

Check
Did I use the correct information?
Did I do what was asked?
Is my answer reasonable?

Name ____________________ Time __________

Power Up K

Use with Lesson 58

Facts Simplify each power or root.

$\sqrt{100} =$	$\sqrt{16} =$	$\sqrt{81} =$	$\sqrt{4} =$	$\sqrt{144} =$
$\sqrt{64} =$	$\sqrt{49} =$	$\sqrt{25} =$	$\sqrt{9} =$	$\sqrt{36} =$
$8^2 =$	$5^2 =$	$3^2 =$	$12^2 =$	$10^2 =$
$7^2 =$	$2^3 =$	$3^3 =$	$10^3 =$	$5^3 =$

Mental Math

a.	**b.**	**c.**	**d.**
e.	**f.**	**g.**	**h.**

Problem Solving

Understand

What information am I given?

What am I asked to find or do?

Plan

How can I use the information I am given?

Which strategy should I try?

Solve

Did I follow the plan?

Did I show my work?

Did I write the answer?

Check

Did I use the correct information?

Did I do what was asked?

Is my answer reasonable?

Name ______________________ Time ____________

Power Up L

Use with Lesson 59

Facts Write the equivalent decimal and percent for each fraction.

Fraction	Decimal	Percent	Fraction	Decimal	Percent
$\frac{1}{2}$			$\frac{1}{8}$		
$\frac{1}{3}$			$\frac{1}{10}$		
$\frac{2}{3}$			$\frac{3}{10}$		
$\frac{1}{4}$			$\frac{9}{10}$		
$\frac{3}{4}$			$\frac{1}{100}$		
$\frac{1}{5}$			$1\frac{1}{2}$		

Mental Math

a.	**b.**	**c.**	**d.**
e.	**f.**	**g.**	**h.**

Problem Solving

Understand
What information am I given?
What am I asked to find or do?

Plan
How can I use the information I am given?
Which strategy should I try?

Solve
Did I follow the plan?
Did I show my work?
Did I write the answer?

Check
Did I use the correct information?
Did I do what was asked?
Is my answer reasonable?

Name ______________________ Time ____________

Power Up M

Use with Lesson 60

Facts

Write the number for each conversion or factor.

1. 2 m = ______ cm
2. 1.5 km = ______ m
3. 2.54 cm = ______ mm
4. 125 cm = ______ m
5. 10 km = ______ m
6. 5000 m = ______ km
7. 50 cm = ______ m
8. 50 cm = ______ mm

9. 2 L = ______ mL
10. 250 mL = ______ L
11. 4 kg = ______ g
12. 2.5 g = ______ mg
13. 500 mg = ______ g
14. 0.5 kg = ______ g

15–16. Two liters of water have a volume of ______ cm^3 and a mass of ___ kg.

	Prefix	Factor
17.	kilo-	
18.	hecto-	
19.	deka-	
	(unit)	
20.	deci-	
21.	centi-	
22.	milli-	

Mental Math

a.	b.	c.	d.
e.	f.	g.	h.

Problem Solving

Understand

What information am I given?

What am I asked to find or do?

Plan

How can I use the information I am given?

Which strategy should I try?

Solve

Did I follow the plan?

Did I show my work?

Did I write the answer?

Check

Did I use the correct information?

Did I do what was asked?

Is my answer reasonable?

Name ______________________ Time __________

Power Up L

Use with Lesson 61

Facts Write the equivalent decimal and percent for each fraction.

Fraction	Decimal	Percent	Fraction	Decimal	Percent
$\frac{1}{2}$			$\frac{1}{8}$		
$\frac{1}{3}$			$\frac{1}{10}$		
$\frac{2}{3}$			$\frac{3}{10}$		
$\frac{1}{4}$			$\frac{9}{10}$		
$\frac{3}{4}$			$\frac{1}{100}$		
$\frac{1}{5}$			$1\frac{1}{2}$		

Mental Math

a.	b.	c.	d.
e.	f.	g.	h.

Problem Solving

Understand

What information am I given?

What am I asked to find or do?

Plan

How can I use the information I am given?

Which strategy should I try?

Solve

Did I follow the plan?

Did I show my work?

Did I write the answer?

Check

Did I use the correct information?

Did I do what was asked?

Is my answer reasonable?

Name ____________________ Time __________

Power Up M

Use with Lesson 62

Facts Write the number for each conversion or factor.

1. 2 m = ______ cm
2. 1.5 km = ______ m
3. 2.54 cm = ______ mm
4. 125 cm = ______ m
5. 10 km = ______ m
6. 5000 m = ______ km
7. 50 cm = ______ m
8. 50 cm = ______ mm
9. 2 L = ______ mL
10. 250 mL = ______ L
11. 4 kg = ______ g
12. 2.5 g = ______ mg
13. 500 mg = ______ g
14. 0.5 kg = ______ g

15–16. Two liters of water have a volume of ______ cm^3 and a mass of ____ kg.

	Prefix	Factor
17.	kilo-	
18.	hecto-	
19.	deka-	
	(unit)	
20.	deci-	
21.	centi-	
22.	milli-	

Mental Math

a.	**b.**	**c.**	**d.**
e.	**f.**	**g.**	**h.**

Problem Solving

Understand
What information am I given?
What am I asked to find or do?

Plan
How can I use the information I am given?
Which strategy should I try?

Solve
Did I follow the plan?
Did I show my work?
Did I write the answer?

Check
Did I use the correct information?
Did I do what was asked?
Is my answer reasonable?

Name ______________________ Time __________

Power Up L

Use with Lesson 63

Facts

Write the equivalent decimal and percent for each fraction.

Fraction	Decimal	Percent	Fraction	Decimal	Percent
$\frac{1}{2}$			$\frac{1}{8}$		
$\frac{1}{3}$			$\frac{1}{10}$		
$\frac{2}{3}$			$\frac{3}{10}$		
$\frac{1}{4}$			$\frac{9}{10}$		
$\frac{3}{4}$			$\frac{1}{100}$		
$\frac{1}{5}$			$1\frac{1}{2}$		

Mental Math

a.	**b.**	**c.**	**d.**
e.	**f.**	**g.**	**h.**

Problem Solving

Understand
What information am I given?
What am I asked to find or do?

Plan
How can I use the information I am given?
Which strategy should I try?

Solve
Did I follow the plan?
Did I show my work?
Did I write the answer?

Check
Did I use the correct information?
Did I do what was asked?
Is my answer reasonable?

Name ____________________ Time ____________

Power Up N

Use with Lesson 64

Facts

Simplify. Reduce the answers if possible.

$3 + 1\frac{2}{3} =$	$3 - 1\frac{2}{3} =$	$3 \times 1\frac{2}{3} =$	$3 \div 1\frac{2}{3} =$
$1\frac{2}{3} + 1\frac{1}{2} =$	$1\frac{2}{3} - 1\frac{1}{2} =$	$1\frac{2}{3} \times 1\frac{1}{2} =$	$1\frac{2}{3} \div 1\frac{1}{2} =$
$2\frac{1}{2} + 1\frac{2}{3} =$	$2\frac{1}{2} - 1\frac{2}{3} =$	$2\frac{1}{2} \times 1\frac{2}{3} =$	$2\frac{1}{2} \div 1\frac{2}{3} =$
$4\frac{1}{2} + 2\frac{1}{4} =$	$4\frac{1}{2} - 2\frac{1}{4} =$	$4\frac{1}{2} \times 2\frac{1}{4} =$	$4\frac{1}{2} \div 2\frac{1}{4} =$

Mental Math

a.	**b.**	**c.**	**d.**
e.	**f.**	**g.**	**h.**

Problem Solving

Understand
What information am I given?
What am I asked to find or do?

Plan
How can I use the information I am given?
Which strategy should I try?

Solve
Did I follow the plan?
Did I show my work?
Did I write the answer?

Check
Did I use the correct information?
Did I do what was asked?
Is my answer reasonable?

Name ______________________ Time __________

Power Up M

Use with Lesson 65

Facts

Write the number for each conversion or factor.

1. 2 m = ______ cm
2. 1.5 km = ______ m
3. 2.54 cm = ______ mm
4. 125 cm = ______ m
5. 10 km = ______ m
6. 5000 m = ______ km
7. 50 cm = ______ m
8. 50 cm = ______ mm
9. 2 L = ______ mL
10. 250 mL = ______ L
11. 4 kg = ______ g
12. 2.5 g = ______ mg
13. 500 mg = ______ g
14. 0.5 kg = ______ g

15–16. Two liters of water have a volume of ______ cm^3 and a mass of ___ kg.

	Prefix	Factor
17.	kilo-	
18.	hecto-	
19.	deka-	
	(unit)	
20.	deci-	
21.	centi-	
22.	milli-	

Mental Math

a.	b.	c.	d.
e.	f.	g.	h.

Problem Solving

Understand

What information am I given?

What am I asked to find or do?

Plan

How can I use the information I am given?

Which strategy should I try?

Solve

Did I follow the plan?

Did I show my work?

Did I write the answer?

Check

Did I use the correct information?

Did I do what was asked?

Is my answer reasonable?

Name ____________________ Time __________

Power Up N

Use with Lesson 66

Facts Simplify. Reduce the answers if possible.

$3 + 1\frac{2}{3} =$	$3 - 1\frac{2}{3} =$	$3 \times 1\frac{2}{3} =$	$3 \div 1\frac{2}{3} =$
$1\frac{2}{3} + 1\frac{1}{2} =$	$1\frac{2}{3} - 1\frac{1}{2} =$	$1\frac{2}{3} \times 1\frac{1}{2} =$	$1\frac{2}{3} \div 1\frac{1}{2} =$
$2\frac{1}{2} + 1\frac{2}{3} =$	$2\frac{1}{2} - 1\frac{2}{3} =$	$2\frac{1}{2} \times 1\frac{2}{3} =$	$2\frac{1}{2} \div 1\frac{2}{3} =$
$4\frac{1}{2} + 2\frac{1}{4} =$	$4\frac{1}{2} - 2\frac{1}{4} =$	$4\frac{1}{2} \times 2\frac{1}{4} =$	$4\frac{1}{2} \div 2\frac{1}{4} =$

Mental Math

a.	**b.**	**c.**	**d.**
e.	**f.**	**g.**	**h.**

Problem Solving

Understand
What information am I given?
What am I asked to find or do?

Plan
How can I use the information I am given?
Which strategy should I try?

Solve
Did I follow the plan?
Did I show my work?
Did I write the answer?

Check
Did I use the correct information?
Did I do what was asked?
Is my answer reasonable?

Name ______________________ Time ____________

Power Up M

Use with Lesson 67

Facts

Write the number for each conversion or factor.

1. 2 m = ______ cm	9. 2 L = ______ mL			Prefix	Factor
2. 1.5 km = ______ m	10. 250 mL = ______ L		17.	kilo-	
3. 2.54 cm = ______ mm	11. 4 kg = ______ g		18.	hecto-	
4. 125 cm = ______ m	12. 2.5 g = ______ mg		19.	deka-	
5. 10 km = ______ m	13. 500 mg = ______ g			(unit)	
6. 5000 m = ______ km	14. 0.5 kg = ______ g		20.	deci-	
7. 50 cm = ______ m	15–16. Two liters of water have a volume of ______ cm^3		21.	centi-	
8. 50 cm = ______ mm	and a mass of ___ kg.		22.	milli-	

Mental Math

a.	**b.**	**c.**	**d.**
e.	**f.**	**g.**	**h.**

Problem Solving

Understand
What information am I given?
What am I asked to find or do?

Plan
How can I use the information I am given?
Which strategy should I try?

Solve
Did I follow the plan?
Did I show my work?
Did I write the answer?

Check
Did I use the correct information?
Did I do what was asked?
Is my answer reasonable?

Name ______________________ Time ____________

Power Up N

Use with Lesson 68

Facts Simplify. Reduce the answers if possible.

$3 + 1\frac{2}{3} =$	$3 - 1\frac{2}{3} =$	$3 \times 1\frac{2}{3} =$	$3 \div 1\frac{2}{3} =$
$1\frac{2}{3} + 1\frac{1}{2} =$	$1\frac{2}{3} - 1\frac{1}{2} =$	$1\frac{2}{3} \times 1\frac{1}{2} =$	$1\frac{2}{3} \div 1\frac{1}{2} =$
$2\frac{1}{2} + 1\frac{2}{3} =$	$2\frac{1}{2} - 1\frac{2}{3} =$	$2\frac{1}{2} \times 1\frac{2}{3} =$	$2\frac{1}{2} \div 1\frac{2}{3} =$
$4\frac{1}{2} + 2\frac{1}{4} =$	$4\frac{1}{2} - 2\frac{1}{4} =$	$4\frac{1}{2} \times 2\frac{1}{4} =$	$4\frac{1}{2} \div 2\frac{1}{4} =$

Mental Math

a.	**b.**	**c.**	**d.**
e.	**f.**	**g.**	**h.**

Problem Solving

Understand
What information am I given?
What am I asked to find or do?

Plan
How can I use the information I am given?
Which strategy should I try?

Solve
Did I follow the plan?
Did I show my work?
Did I write the answer?

Check
Did I use the correct information?
Did I do what was asked?
Is my answer reasonable?

Name ______________________ Time __________

Power Up M

Use with Lesson 69

Facts Write the number for each conversion or factor.

1. 2 m = ______ cm	9. 2 L = ______ mL
2. 1.5 km = ______ m	10. 250 mL = ______ L
3. 2.54 cm = ______ mm	11. 4 kg = ______ g
4. 125 cm = ______ m	12. 2.5 g = ______ mg
5. 10 km = ______ m	13. 500 mg = ______ g
6. 5000 m = ______ km	14. 0.5 kg = ______ g
7. 50 cm = ______ m	15–16. Two liters of water have a volume of ______ cm^3 and a mass of ____ kg.
8. 50 cm = ______ mm	

	Prefix	Factor
17.	kilo-	
18.	hecto-	
19.	deka-	
	(unit)	
20.	deci-	
21.	centi-	
22.	milli-	

Mental Math

a.	**b.**	**c.**	**d.**
e.	**f.**	**g.**	**h.**

Problem Solving

Understand
What information am I given?
What am I asked to find or do?

Plan
How can I use the information I am given?
Which strategy should I try?

Solve
Did I follow the plan?
Did I show my work?
Did I write the answer?

Check
Did I use the correct information?
Did I do what was asked?
Is my answer reasonable?

Name ______________________ Time __________

Power Up N

Use with Lesson 70

Facts

Simplify. Reduce the answers if possible.

$3 + 1\frac{2}{3} =$	$3 - 1\frac{2}{3} =$	$3 \times 1\frac{2}{3} =$	$3 \div 1\frac{2}{3} =$
$1\frac{2}{3} + 1\frac{1}{2} =$	$1\frac{2}{3} - 1\frac{1}{2} =$	$1\frac{2}{3} \times 1\frac{1}{2} =$	$1\frac{2}{3} \div 1\frac{1}{2} =$
$2\frac{1}{2} + 1\frac{2}{3} =$	$2\frac{1}{2} - 1\frac{2}{3} =$	$2\frac{1}{2} \times 1\frac{2}{3} =$	$2\frac{1}{2} \div 1\frac{2}{3} =$
$4\frac{1}{2} + 2\frac{1}{4} =$	$4\frac{1}{2} - 2\frac{1}{4} =$	$4\frac{1}{2} \times 2\frac{1}{4} =$	$4\frac{1}{2} \div 2\frac{1}{4} =$

Mental Math

a.	**b.**	**c.**	**d.**
e.	**f.**	**g.**	**h.**

Problem Solving

Understand
What information am I given?
What am I asked to find or do?

Plan
How can I use the information I am given?
Which strategy should I try?

Solve
Did I follow the plan?
Did I show my work?
Did I write the answer?

Check
Did I use the correct information?
Did I do what was asked?
Is my answer reasonable?

Name ______________________ Time ____________

Power Up O

Use with Lesson 71

Facts

Select from the words below to describe each figure.

1. 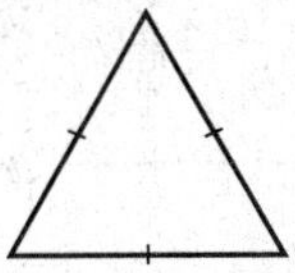	2. 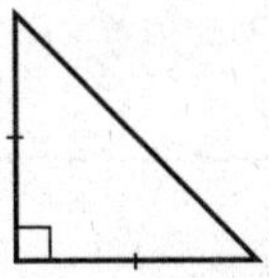	3. 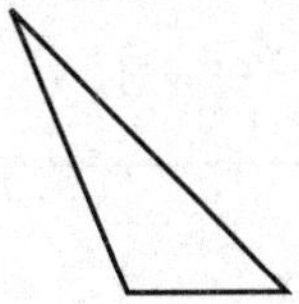	4. 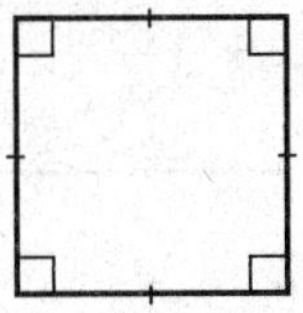
5. 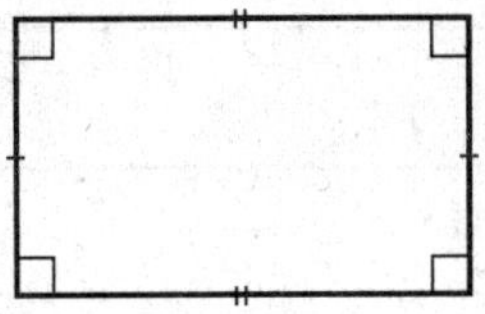	6. 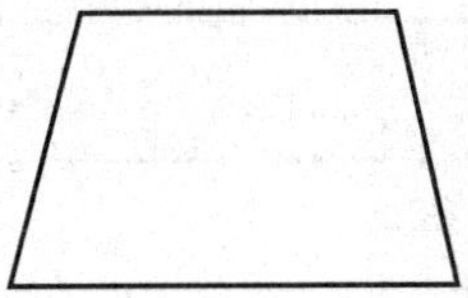	7. 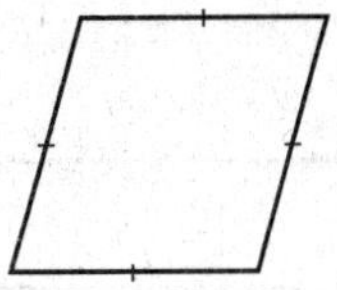	8. 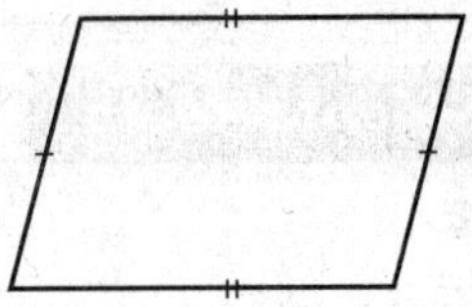
kite trapezoid parallelogram	rectangle rhombus square	isosceles triangle scalene triangle equilateral triangle	right triangle acute triangle obtuse triangle

Mental Math

a.	**b.**	**c.**	**d.**
e.	**f.**	**g.**	**h.**

Problem Solving

Understand

What information am I given?

What am I asked to find or do?

Plan

How can I use the information I am given?

Which strategy should I try?

Solve

Did I follow the plan?

Did I show my work?

Did I write the answer?

Check

Did I use the correct information?

Did I do what was asked?

Is my answer reasonable?

Name ______________________________ Time ______________

Power Up N

Use with Lesson 72

Facts Simplify. Reduce the answers if possible.

$3 + 1\frac{2}{3} =$	$3 - 1\frac{2}{3} =$	$3 \times 1\frac{2}{3} =$	$3 \div 1\frac{2}{3} =$
$1\frac{2}{3} + 1\frac{1}{2} =$	$1\frac{2}{3} - 1\frac{1}{2} =$	$1\frac{2}{3} \times 1\frac{1}{2} =$	$1\frac{2}{3} \div 1\frac{1}{2} =$
$2\frac{1}{2} + 1\frac{2}{3} =$	$2\frac{1}{2} - 1\frac{2}{3} =$	$2\frac{1}{2} \times 1\frac{2}{3} =$	$2\frac{1}{2} \div 1\frac{2}{3} =$
$4\frac{1}{2} + 2\frac{1}{4} =$	$4\frac{1}{2} - 2\frac{1}{4} =$	$4\frac{1}{2} \times 2\frac{1}{4} =$	$4\frac{1}{2} \div 2\frac{1}{4} =$

Mental Math

a.	**b.**	**c.**	**d.**
e.	**f.**	**g.**	**h.**

Problem Solving

Understand
What information am I given?
What am I asked to find or do?

Plan
How can I use the information I am given?
Which strategy should I try?

Solve
Did I follow the plan?
Did I show my work?
Did I write the answer?

Check
Did I use the correct information?
Did I do what was asked?
Is my answer reasonable?

Name ______________________ Time __________

Power Up O

Use with Lesson 73

Facts

Select from the words below to describe each figure.

1. 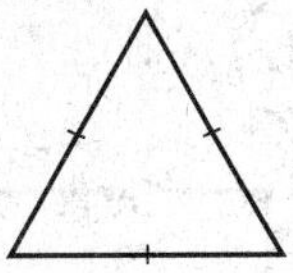	2. 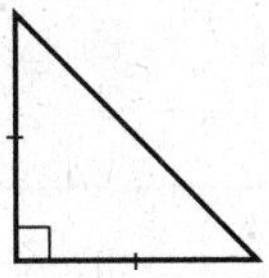	3. 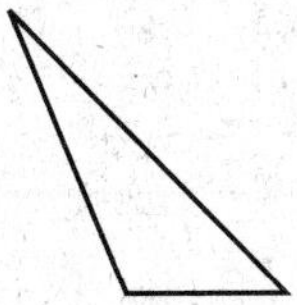	4. 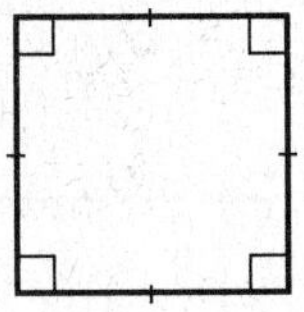
5. 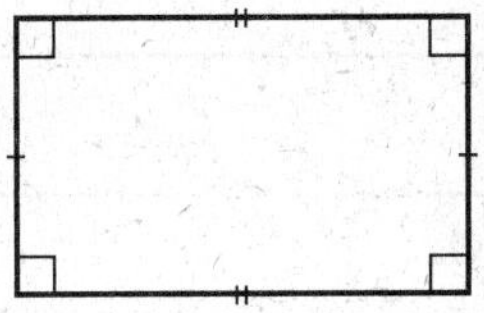	6. 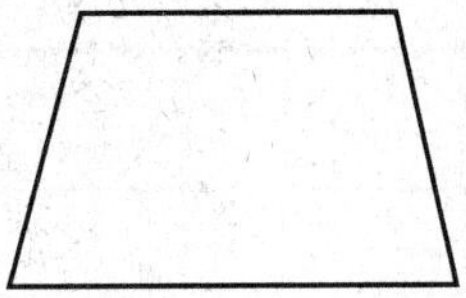	7. 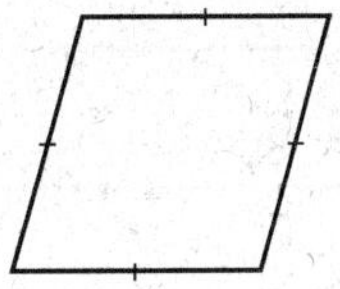	8. 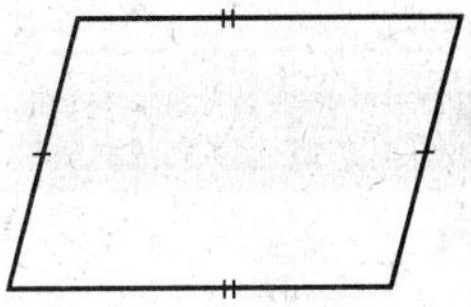

kite	rectangle	isosceles triangle	right triangle
trapezoid	rhombus	scalene triangle	acute triangle
parallelogram	square	equilateral triangle	obtuse triangle

Mental Math

a.	**b.**	**c.**	**d.**
e.	**f.**	**g.**	**h.**

Problem Solving

Understand
What information am I given?
What am I asked to find or do?

Plan
How can I use the information I am given?
Which strategy should I try?

Solve
Did I follow the plan?
Did I show my work?
Did I write the answer?

Check
Did I use the correct information?
Did I do what was asked?
Is my answer reasonable?

Name ______________________ Time ____________

Power Up N

Use with Lesson 74

Facts

Simplify. Reduce the answers if possible.

$3 + 1\frac{2}{3} =$	$3 - 1\frac{2}{3} =$	$3 \times 1\frac{2}{3} =$	$3 \div 1\frac{2}{3} =$
$1\frac{2}{3} + 1\frac{1}{2} =$	$1\frac{2}{3} - 1\frac{1}{2} =$	$1\frac{2}{3} \times 1\frac{1}{2} =$	$1\frac{2}{3} \div 1\frac{1}{2} =$
$2\frac{1}{2} + 1\frac{2}{3} =$	$2\frac{1}{2} - 1\frac{2}{3} =$	$2\frac{1}{2} \times 1\frac{2}{3} =$	$2\frac{1}{2} \div 1\frac{2}{3} =$
$4\frac{1}{2} + 2\frac{1}{4} =$	$4\frac{1}{2} - 2\frac{1}{4} =$	$4\frac{1}{2} \times 2\frac{1}{4} =$	$4\frac{1}{2} \div 2\frac{1}{4} =$

Mental Math

a.	**b.**	**c.**	**d.**
e.	**f.**	**g.**	**h.**

Problem Solving

Understand
What information am I given?
What am I asked to find or do?

Plan
How can I use the information I am given?
Which strategy should I try?

Solve
Did I follow the plan?
Did I show my work?
Did I write the answer?

Check
Did I use the correct information?
Did I do what was asked?
Is my answer reasonable?

Name ______________________ Time ____________

Power Up O

Use with Lesson 75

Facts

Select from the words below to describe each figure.

1. 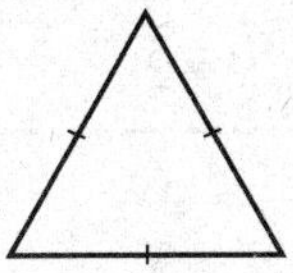

2. 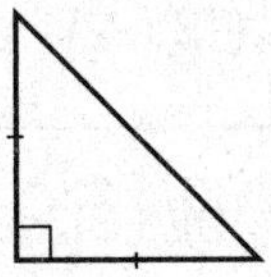

3. 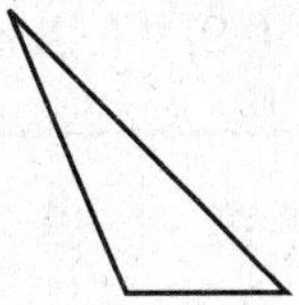

4. 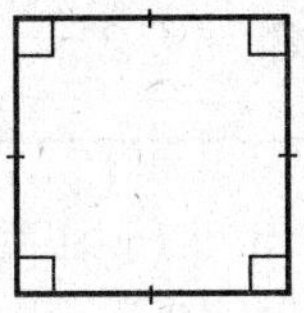

5. 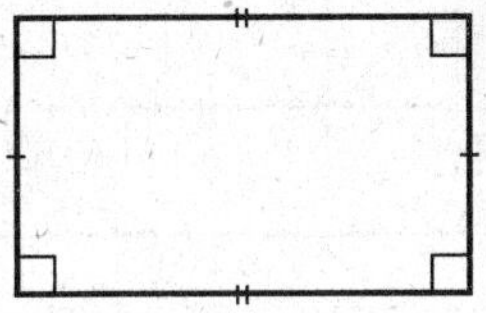

6. 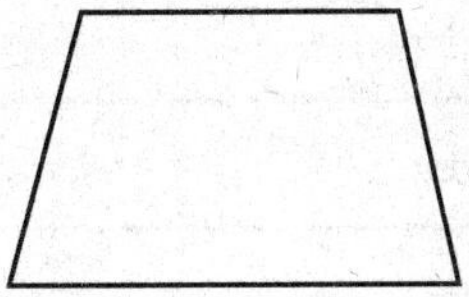

7. 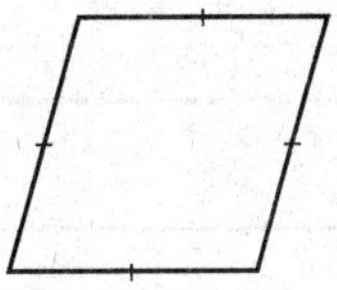

8. 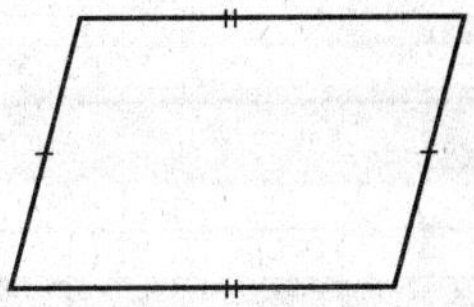

kite trapezoid parallelogram	rectangle rhombus square	isosceles triangle scalene triangle equilateral triangle	right triangle acute triangle obtuse triangle

Mental Math

a.	**b.**	**c.**	**d.**
e.	**f.**	**g.**	**h.**

Problem Solving

Understand
What information am I given?
What am I asked to find or do?

Plan
How can I use the information I am given?
Which strategy should I try?

Solve
Did I follow the plan?
Did I show my work?
Did I write the answer?

Check
Did I use the correct information?
Did I do what was asked?
Is my answer reasonable?

Name ____________________ Time __________

Power Up P

Use with Lesson 76

Facts Simplify.

$(-8) + (-2) =$	$(-8) - (-2) =$	$(-8)(-2) =$	$\frac{-8}{-2} =$
$(-9) + (+3) =$	$(-9) - (+3) =$	$(-9)(+3) =$	$\frac{-9}{+3} =$
$12 + (-2) =$	$12 - (-2) =$	$(12)(-2) =$	$\frac{12}{-2} =$
$(-4) + (-3) + (-2) =$	$(-4) - (-3) - (-2) =$	$(-4)(-3)(-2) =$	$\frac{(-4)(-3)}{(-2)} =$

Mental Math

a.	**b.**	**c.**	**d.**
e.	**f.**	**g.**	**h.**

Problem Solving

Understand
What information am I given?
What am I asked to find or do?

Plan
How can I use the information I am given?
Which strategy should I try?

Solve
Did I follow the plan?
Did I show my work?
Did I write the answer?

Check
Did I use the correct information?
Did I do what was asked?
Is my answer reasonable?

Name ______________________ Time ____________

Power Up P

Use with Lesson 77

Facts

Simplify.

$(-8) + (-2) =$	$(-8) - (-2) =$	$(-8)(-2) =$	$\frac{-8}{-2} =$
$(-9) + (+3) =$	$(-9) - (+3) =$	$(-9)(+3) =$	$\frac{-9}{+3} =$
$12 + (-2) =$	$12 - (-2) =$	$(12)(-2) =$	$\frac{12}{-2} =$
$(-4) + (-3) + (-2) =$	$(-4) - (-3) - (-2) =$	$(-4)(-3)(-2) =$	$\frac{(-4)(-3)}{(-2)} =$

Mental Math

a.	**b.**	**c.**	**d.**
e.	**f.**	**g.**	**h.**

Problem Solving

Understand

What information am I given?

What am I asked to find or do?

Plan

How can I use the information I am given?

Which strategy should I try?

Solve

Did I follow the plan?

Did I show my work?

Did I write the answer?

Check

Did I use the correct information?

Did I do what was asked?

Is my answer reasonable?

Name ______________________ Time ____________

Power Up O

Use with Lesson 78

Facts

Select from the words below to describe each figure.

1. 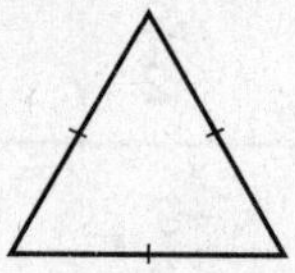	2. 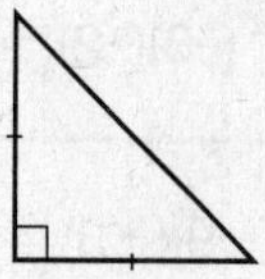	3. 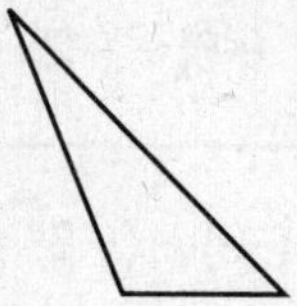	4. 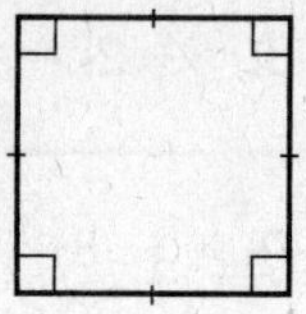
5. 	6. 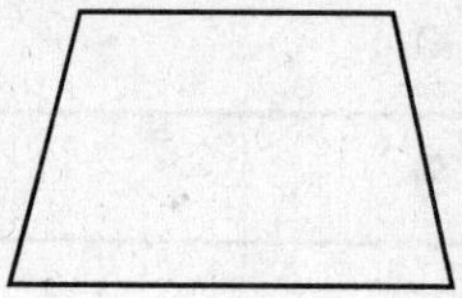	7. 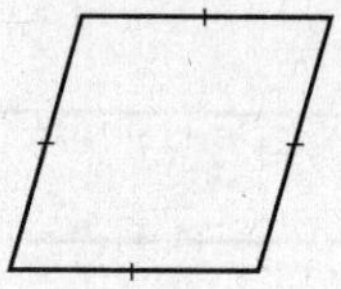	8. 
kite trapezoid parallelogram	rectangle rhombus square	isosceles triangle scalene triangle equilateral triangle	right triangle acute triangle obtuse triangle

Mental Math

a.	**b.**	**c.**	**d.**
e.	**f.**	**g.**	**h.**

Problem Solving

Understand

What information am I given?

What am I asked to find or do?

Plan

How can I use the information I am given?

Which strategy should I try?

Solve

Did I follow the plan?

Did I show my work?

Did I write the answer?

Check

Did I use the correct information?

Did I do what was asked?

Is my answer reasonable?

Name ______________________ Time ____________

Power Up P

Use with Lesson 79

Facts Simplify.

$(-8) + (-2) =$	$(-8) - (-2) =$	$(-8)(-2) =$	$\frac{-8}{-2} =$
$(-9) + (+3) =$	$(-9) - (+3) =$	$(-9)(+3) =$	$\frac{-9}{+3} =$
$12 + (-2) =$	$12 - (-2) =$	$(12)(-2) =$	$\frac{12}{-2} =$
$(-4) + (-3) + (-2) =$	$(-4) - (-3) - (-2) =$	$(-4)(-3)(-2) =$	$\frac{(-4)(-3)}{(-2)} =$

Mental Math

a.	**b.**	**c.**	**d.**
e.	**f.**	**g.**	**h.**

Problem Solving

Understand

What information am I given?

What am I asked to find or do?

Plan

How can I use the information I am given?

Which strategy should I try?

Solve

Did I follow the plan?

Did I show my work?

Did I write the answer?

Check

Did I use the correct information?

Did I do what was asked?

Is my answer reasonable?

Name ____________________ Time __________

Power Up O

Use with Lesson 80

Facts

Select from the words below to describe each figure.

1. 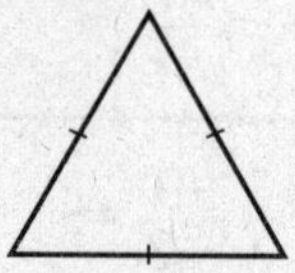	2. 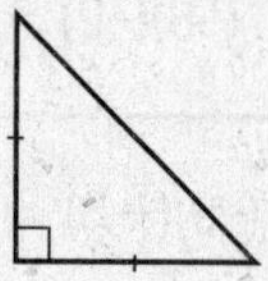	3. 	4. 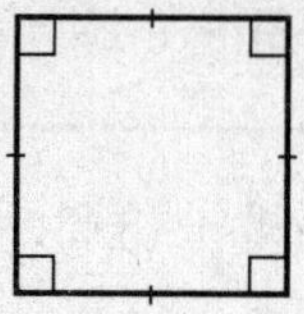
5. 	6. 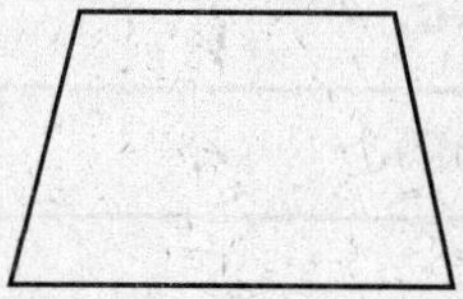	7. 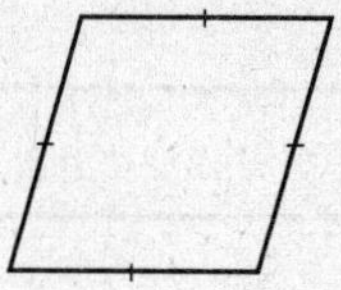	8. 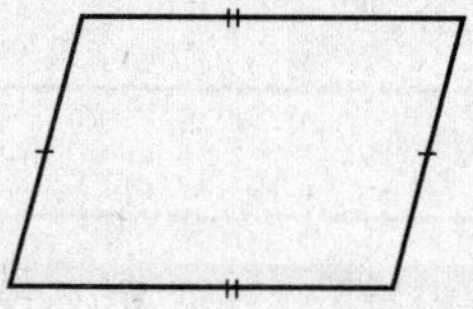
kite trapezoid parallelogram	rectangle rhombus square	isosceles triangle scalene triangle equilateral triangle	right triangle acute triangle obtuse triangle

Mental Math

a.	**b.**	**c.**	**d.**
e.	**f.**	**g.**	**h.**

Problem Solving

Understand

What information am I given?
What am I asked to find or do?

Plan

How can I use the information I am given?
Which strategy should I try?

Solve

Did I follow the plan?
Did I show my work?
Did I write the answer?

Check

Did I use the correct information?
Did I do what was asked?
Is my answer reasonable?

Name ______________________ Time ____________

Power Up P

Use with Lesson 81

Facts Simplify.

$(-8) + (-2) =$	$(-8) - (-2) =$	$(-8)(-2) =$	$\frac{-8}{-2} =$
$(-9) + (+3) =$	$(-9) - (+3) =$	$(-9)(+3) =$	$\frac{-9}{+3} =$
$12 + (-2) =$	$12 - (-2) =$	$(12)(-2) =$	$\frac{12}{-2} =$
$(-4) + (-3) + (-2) =$	$(-4) - (-3) - (-2) =$	$(-4)(-3)(-2) =$	$\frac{(-4)(-3)}{(-2)} =$

Mental Math

a.	**b.**	**c.**	**d.**
e.	**f.**	**g.**	**h.**

Problem Solving

Understand

What information am I given?

What am I asked to find or do?

Plan

How can I use the information I am given?

Which strategy should I try?

Solve

Did I follow the plan?

Did I show my work?

Did I write the answer?

Check

Did I use the correct information?

Did I do what was asked?

Is my answer reasonable?

Name ______________________ Time ____________

Power Up Q

Use with Lesson 82

Facts Write the equivalent decimal and fraction for each percent.

Percent	Decimal	Fraction	Percent	Decimal	Fraction
10%			$33\frac{1}{3}\%$		
90%			20%		
5%			75%		
$12\frac{1}{2}\%$			$66\frac{2}{3}\%$		
50%			1%		
25%			250%		

Mental Math

a.	b.	c.	d.
e.	f.	g.	h.

Problem Solving

Understand

What information am I given?

What am I asked to find or do?

Plan

How can I use the information I am given?

Which strategy should I try?

Solve

Did I follow the plan?

Did I show my work?

Did I write the answer?

Check

Did I use the correct information?

Did I do what was asked?

Is my answer reasonable?

Name ______________________ Time ____________

Power Up R

Use with Lesson 83

Facts

Find the area of each figure. Angles that look like right angles are right angles.

1.
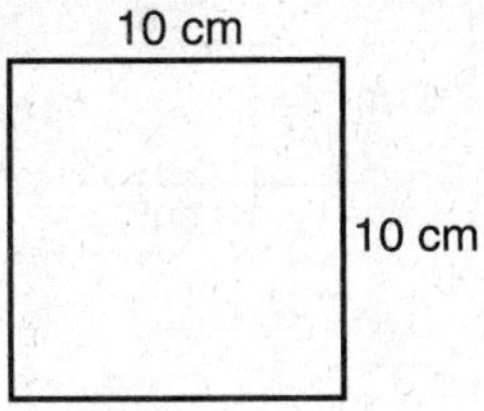

______

2.
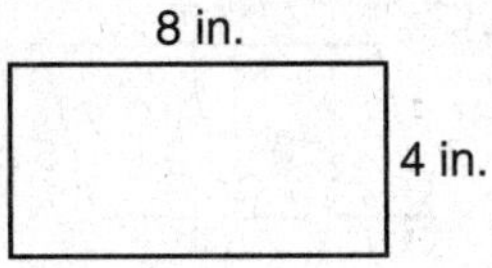

______

3.
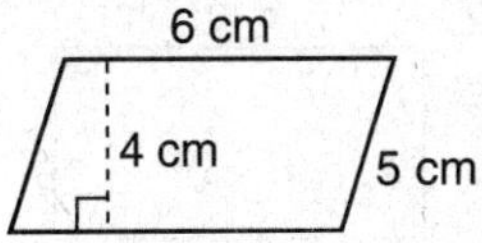

______

4.
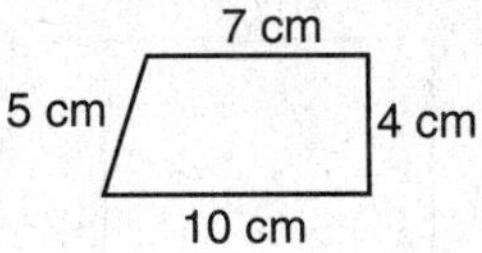

______

5.
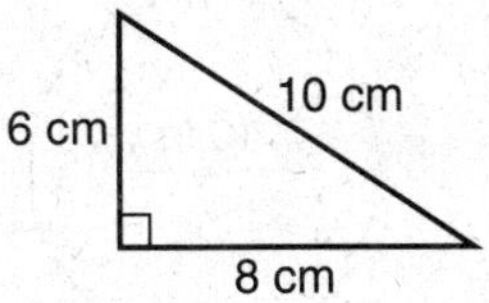

______

6.
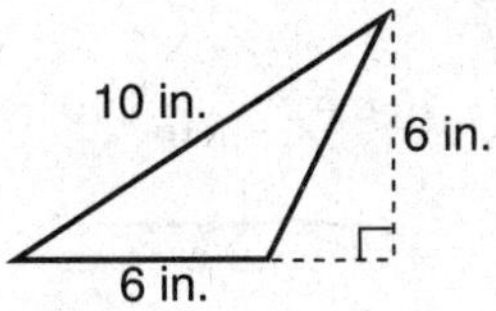

______

7.
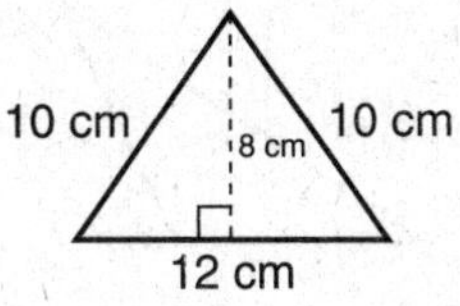

______

8.
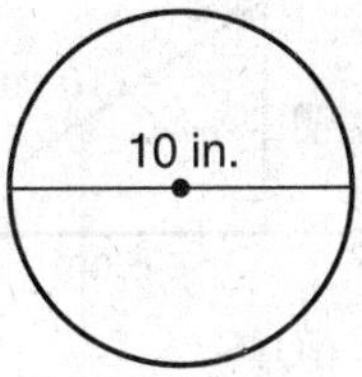

Leave π as π.

______

Mental Math

a.	b.	c.	d.
e.	f.	g.	h.

Problem Solving

Understand
What information am I given?
What am I asked to find or do?

Plan
How can I use the information I am given?
Which strategy should I try?

Solve
Did I follow the plan?
Did I show my work?
Did I write the answer?

Check
Did I use the correct information?
Did I do what was asked?
Is my answer reasonable?

Name ____________________ Time __________

Power Up R

Use with Lesson 84

Facts

Find the area of each figure. Angles that look like right angles are right angles.

1.
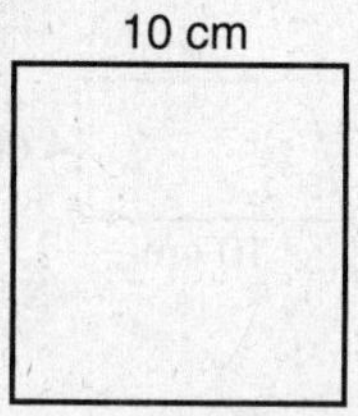

2.
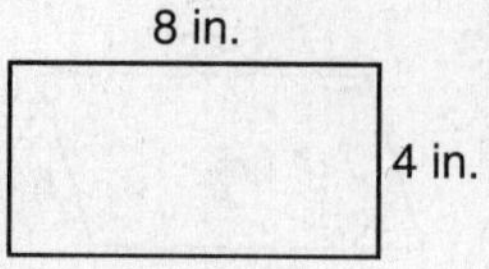

3.
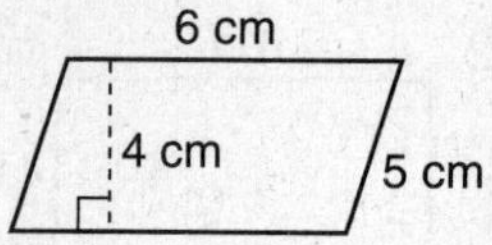

4.
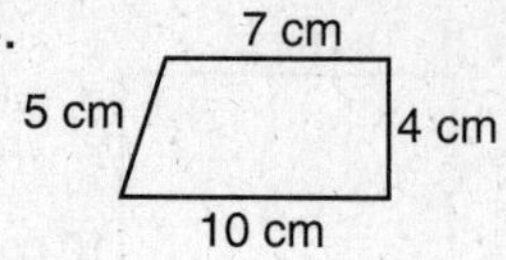

5.
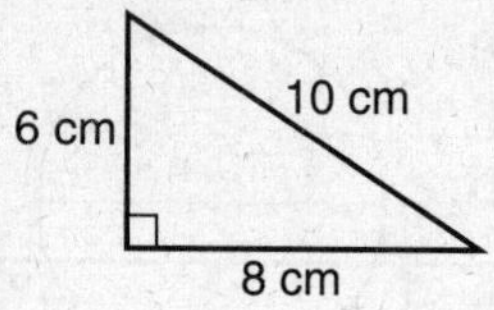

6.
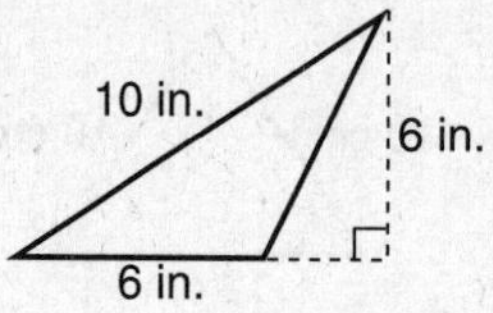

7.
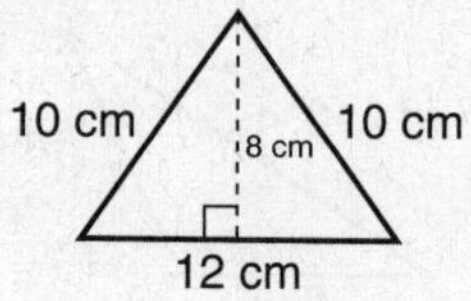

8.
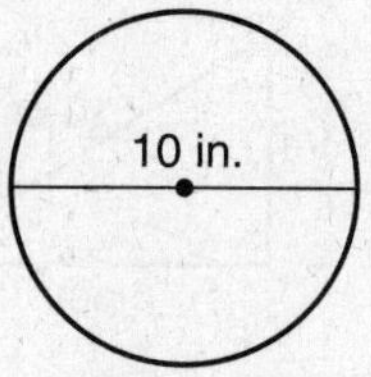

Leave π as π.

Mental Math

a.	b.	c.	d.
e.	f.	g.	h.

Problem Solving

Understand

What information am I given?
What am I asked to find or do?

Plan

How can I use the information I am given?
Which strategy should I try?

Solve

Did I follow the plan?
Did I show my work?
Did I write the answer?

Check

Did I use the correct information?
Did I do what was asked?
Is my answer reasonable?

Name ______________________ Time ____________

Power Up **P**

Use with Lesson 85

Facts Simplify.

$(-8) + (-2) =$	$(-8) - (-2) =$	$(-8)(-2) =$	$\frac{-8}{-2} =$
$(-9) + (+3) =$	$(-9) - (+3) =$	$(-9)(+3) =$	$\frac{-9}{+3} =$
$12 + (-2) =$	$12 - (-2) =$	$(12)(-2) =$	$\frac{12}{-2} =$
$(-4) + (-3) + (-2) =$	$(-4) - (-3) - (-2) =$	$(-4)(-3)(-2) =$	$\frac{(-4)(-3)}{(-2)} =$

Mental Math

a.	**b.**	**c.**	**d.**
e.	**f.**	**g.**	**h.**

Problem Solving

Understand

What information am I given?

What am I asked to find or do?

Plan

How can I use the information I am given?

Which strategy should I try?

Solve

Did I follow the plan?

Did I show my work?

Did I write the answer?

Check

Did I use the correct information?

Did I do what was asked?

Is my answer reasonable?

Name ______________________ Time ____________

Power Up S

Use with Lesson 86

Facts

Write each number in scientific notation.

186,000 =	0.0005 =	30,500,000 =
2.5 billion =	12 million =	$\frac{1}{1{,}000{,}000}$ =

Write each number in standard form.

1×10^6 =	1×10^{-6} =	2.4×10^4 =
5×10^{-4} =	4.75×10^5 =	2.5×10^{-3} =

Mental Math

a.	**b.**	**c.**	**d.**
e.	**f.**	**g.**	**h.**

Problem Solving

Understand

What information am I given?

What am I asked to find or do?

Plan

How can I use the information I am given?

Which strategy should I try?

Solve

Did I follow the plan?

Did I show my work?

Did I write the answer?

Check

Did I use the correct information?

Did I do what was asked?

Is my answer reasonable?

Name ______________________ Time __________

Power Up R

Use with Lesson 87

Facts

Find the area of each figure. Angles that look like right angles are right angles.

1.
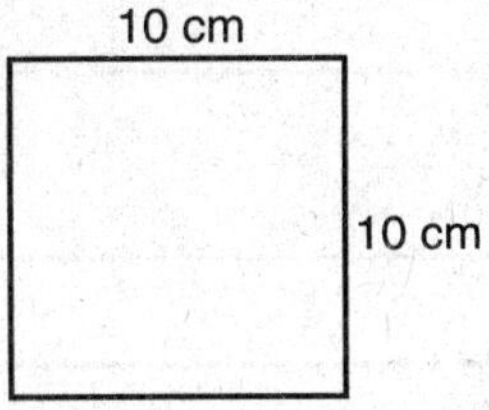

2.
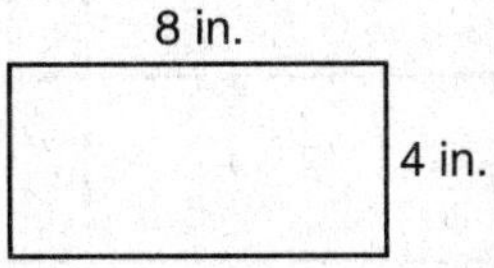

3.
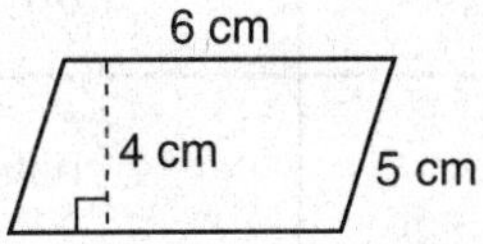

4.
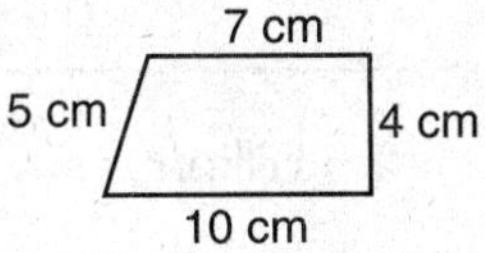

5.
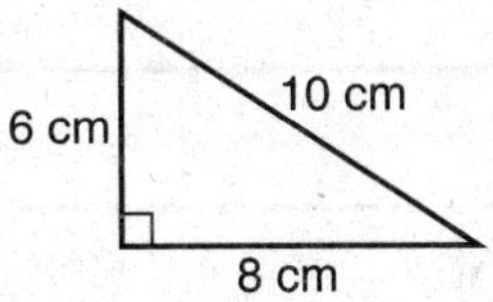

6.
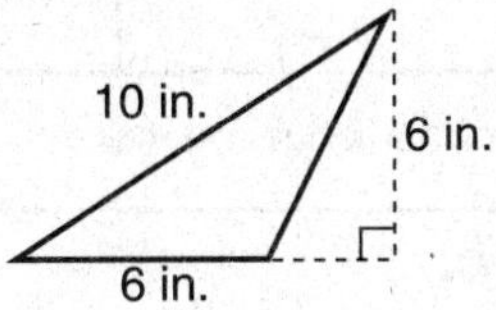

7.
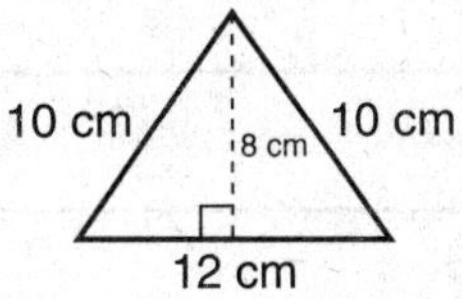

8.
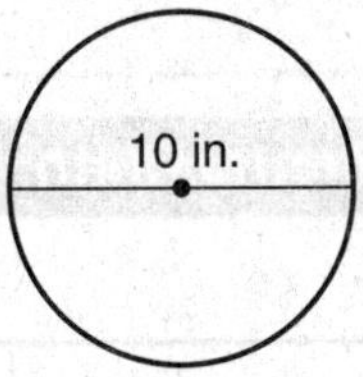

Leave π as π.

Mental Math

a.	b.	c.	d.
e.	f.	g.	h.

Problem Solving

Understand
What information am I given?
What am I asked to find or do?

Plan
How can I use the information I am given?
Which strategy should I try?

Solve
Did I follow the plan?
Did I show my work?
Did I write the answer?

Check
Did I use the correct information?
Did I do what was asked?
Is my answer reasonable?

Name ____________________ Time __________

Power Up S

Use with Lesson 88

Facts

Write each number in scientific notation.

186,000 =	0.0005 =	30,500,000 =
2.5 billion =	12 million =	$\frac{1}{1{,}000{,}000}$ =

Write each number in standard form.

1×10^6 =	1×10^{-6} =	2.4×10^4 =
5×10^{-4} =	4.75×10^5 =	2.5×10^{-3} =

Mental Math

a.	**b.**	**c.**	**d.**
e.	**f.**	**g.**	**h.**

Problem Solving

Understand

What information am I given?

What am I asked to find or do?

Plan

How can I use the information I am given?

Which strategy should I try?

Solve

Did I follow the plan?

Did I show my work?

Did I write the answer?

Check

Did I use the correct information?

Did I do what was asked?

Is my answer reasonable?

Name ____________________ Time ____________

Power Up R

Use with Lesson 89

Facts

Find the area of each figure. Angles that look like right angles are right angles.

1.
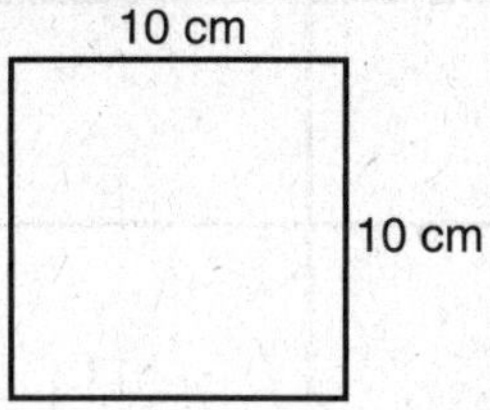

2.
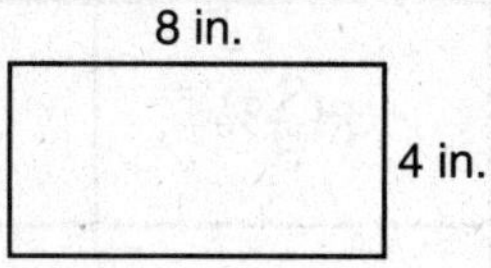

3.
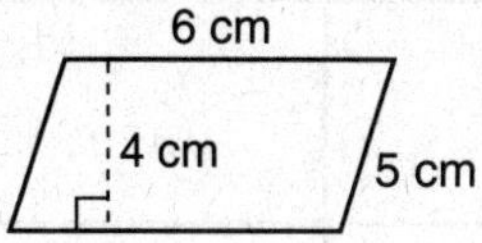

4.
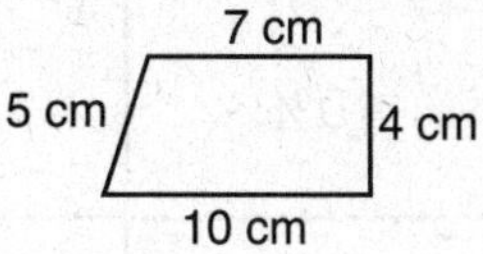

5.
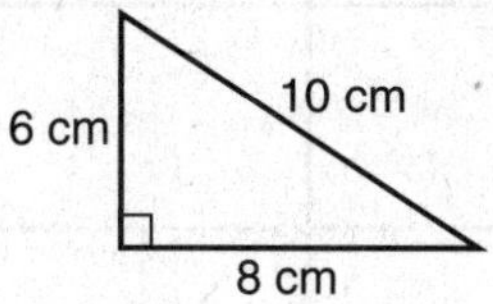

6.
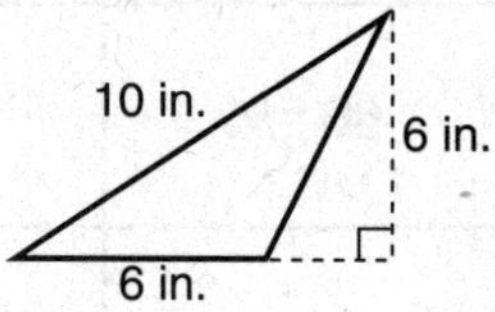

7.
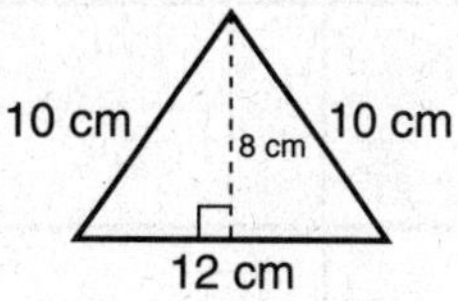

8.
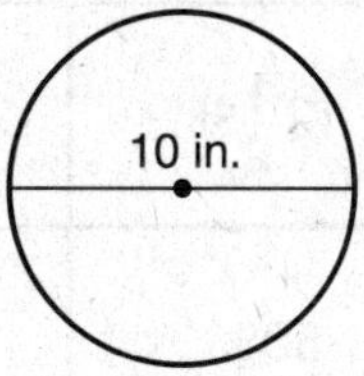

Leave π as π.

Mental Math

a.	**b.**	**c.**	**d.**
e.	**f.**	**g.**	**h.**

Problem Solving

Understand
What information am I given?
What am I asked to find or do?

Plan
How can I use the information I am given?
Which strategy should I try?

Solve
Did I follow the plan?
Did I show my work?
Did I write the answer?

Check
Did I use the correct information?
Did I do what was asked?
Is my answer reasonable?

Name ______________________ Time __________

Power Up Q

Use with Lesson 90

Facts

Write the equivalent decimal and fraction for each percent.

Percent	Decimal	Fraction	Percent	Decimal	Fraction
10%			$33\frac{1}{3}$%		
90%			20%		
5%			75%		
$12\frac{1}{2}$%			$66\frac{2}{3}$%		
50%			1%		
25%			250%		

Mental Math

a.	b.	c.	d.
e.	f.	g.	h.

Problem Solving

Understand
What information am I given?
What am I asked to find or do?

Plan
How can I use the information I am given?
Which strategy should I try?

Solve
Did I follow the plan?
Did I show my work?
Did I write the answer?

Check
Did I use the correct information?
Did I do what was asked?
Is my answer reasonable?

Name ______________________ Time __________

Power Up S

Use with Lesson 91

Facts

Write each number in scientific notation.

186,000 =	0.0005 =	30,500,000 =
2.5 billion =	12 million =	$\frac{1}{1,000,000}$ =

Write each number in standard form.

1×10^6 =	1×10^{-6} =	2.4×10^4 =
5×10^{-4} =	4.75×10^5 =	2.5×10^{-3} =

Mental Math

a.	**b.**	**c.**	**d.**
e.	**f.**	**g.**	**h.**

Problem Solving

Understand

What information am I given?

What am I asked to find or do?

Plan

How can I use the information I am given?

Which strategy should I try?

Solve

Did I follow the plan?

Did I show my work?

Did I write the answer?

Check

Did I use the correct information?

Did I do what was asked?

Is my answer reasonable?

Name ________________________ Time __________

Power Up T

Use with Lesson 92

Facts Simplify.

$6 + 6 \times 6 - 6 \div 6 =$	$3^2 + \sqrt{4} + 5(6) - 7 + 8 =$
$4 + 2(3 + 5) - 6 \div 2 =$	$2 + 2[3 + 4(7 - 5)] =$
$\sqrt{1^3 + 2^3 + 3^3} =$	$\dfrac{4 + 3(7 - 5)}{6 - (5 - 4)} =$
$(-3)(-3) + (-3) - (-3) =$	$\dfrac{3(-3) - (-3)(-3)}{(-3) - (3)(-3)} =$

Mental Math

a.	**b.**	**c.**	**d.**
e.	**f.**	**g.**	**h.**

Problem Solving

Understand

What information am I given?

What am I asked to find or do?

Plan

How can I use the information I am given?

Which strategy should I try?

Solve

Did I follow the plan?

Did I show my work?

Did I write the answer?

Check

Did I use the correct information?

Did I do what was asked?

Is my answer reasonable?

Name ______________________ Time __________

Power Up T

Use with Lesson 93

Facts Simplify.

$6 + 6 \times 6 - 6 \div 6 =$	$3^2 + \sqrt{4} + 5(6) - 7 + 8 =$
$4 + 2(3 + 5) - 6 \div 2 =$	$2 + 2[3 + 4(7 - 5)] =$
$\sqrt{1^3 + 2^3 + 3^3} =$	$\dfrac{4 + 3(7 - 5)}{6 - (5 - 4)} =$
$(-3)(-3) + (-3) - (-3) =$	$\dfrac{3(-3) - (-3)(-3)}{(-3) - (3)(-3)} =$

Mental Math

a.	**b.**	**c.**	**d.**
e.	**f.**	**g.**	**h.**

Problem Solving

Understand

What information am I given?

What am I asked to find or do?

Plan

How can I use the information I am given?

Which strategy should I try?

Solve

Did I follow the plan?

Did I show my work?

Did I write the answer?

Check

Did I use the correct information?

Did I do what was asked?

Is my answer reasonable?

Name ______________________ Time ____________

Power Up A

Use with Lesson 94

Facts Multiply.

9 × 8	8 × 2	10 × 10	6 × 3	4 × 2	5 × 5	9 × 9	6 × 4	9 × 6	7 × 3
9 × 3	6 × 5	0 × 0	7 × 6	8 × 8	7 × 4	5 × 3	9 × 7	2 × 2	8 × 6
7 × 7	6 × 2	4 × 3	8 × 5	4 × 4	3 × 2	*n* × 0	8 × 4	6 × 6	9 × 2
8 × 3	5 × 4	*n* × 1	7 × 2	9 × 5	8 × 7	3 × 3	9 × 4	5 × 2	7 × 5

Mental Math

a.	**b.**	**c.**	**d.**
e.	**f.**	**g.**	**h.**

Problem Solving

Understand
What information am I given?
What am I asked to find or do?

Plan
How can I use the information I am given?
Which strategy should I try?

Solve
Did I follow the plan?
Did I show my work?
Did I write the answer?

Check
Did I use the correct information?
Did I do what was asked?
Is my answer reasonable?

Name ____________________ Time ________

Power Up T

Use with Lesson 95

Facts Simplify.

$6 + 6 \times 6 - 6 \div 6 =$	$3^2 + \sqrt{4} + 5(6) - 7 + 8 =$
$4 + 2(3 + 5) - 6 \div 2 =$	$2 + 2[3 + 4(7 - 5)] =$
$\sqrt{1^3 + 2^3 + 3^3} =$	$\dfrac{4 + 3(7 - 5)}{6 - (5 - 4)} =$
$(-3)(-3) + (-3) - (-3) =$	$\dfrac{3(-3) - (-3)(-3)}{(-3) - (3)(-3)} =$

Mental Math

a.	**b.**	**c.**	**d.**
e.	**f.**	**g.**	**h.**

Problem Solving

Understand

What information am I given?

What am I asked to find or do?

Plan

How can I use the information I am given?

Which strategy should I try?

Solve

Did I follow the plan?

Did I show my work?

Did I write the answer?

Check

Did I use the correct information?

Did I do what was asked?

Is my answer reasonable?

Name ____________________ Time __________

Power Up U

Use with Lesson 96

Facts Complete each step to solve each equation.

$2x + 5 = 45$ $2x =$ $x =$	$3y + 4 = 22$ $3y =$ $y =$	$5n - 3 = 12$ $5n =$ $n =$	$3m - 7 = 14$ $3m =$ $m =$
$15 = 3a - 6$ $= 3a$ $= a$	$24 = 2w + 6$ $= 2w$ $= w$	$-2x + 9 = 23$ $-2x =$ $x =$	$20 - 3y = 2$ $-3y =$ $y =$
$\frac{1}{2}m + 6 = 18$ $\frac{1}{2}m =$ $m =$	$\frac{3}{4}n - 12 = 12$ $\frac{3}{4}n =$ $n =$	$3y + 1.5 = 6$ $3y =$ $y =$	$0.5w - 1.5 = 4.5$ $0.5w =$ $w =$

Mental Math

a.	**b.**	**c.**	**d.**
e.	**f.**	**g.**	**h.**

Problem Solving

Understand
What information am I given?
What am I asked to find or do?

Plan
How can I use the information I am given?
Which strategy should I try?

Solve
Did I follow the plan?
Did I show my work?
Did I write the answer?

Check
Did I use the correct information?
Did I do what was asked?
Is my answer reasonable?

Name ______________________ Time __________

Power Up T

Use with Lesson 97

Facts

Simplify.

$6 + 6 \times 6 - 6 \div 6 =$	$3^2 + \sqrt{4} + 5(6) - 7 + 8 =$
$4 + 2(3 + 5) - 6 \div 2 =$	$2 + 2[3 + 4(7 - 5)] =$
$\sqrt{1^3 + 2^3 + 3^3} =$	$\frac{4 + 3(7 - 5)}{6 - (5 - 4)} =$
$(-3)(-3) + (-3) - (-3) =$	$\frac{3(-3) - (-3)(-3)}{(-3) - (3)(-3)} =$

Mental Math

a.	b.	c.	d.
e.	f.	g.	h.

Problem Solving

Understand

What information am I given?

What am I asked to find or do?

Plan

How can I use the information I am given?

Which strategy should I try?

Solve

Did I follow the plan?

Did I show my work?

Did I write the answer?

Check

Did I use the correct information?

Did I do what was asked?

Is my answer reasonable?

Name ______________________ Time __________

Power Up U

Use with Lesson 98

Facts

Complete each step to solve each equation.

$2x + 5 = 45$ $2x =$ $x =$	$3y + 4 = 22$ $3y =$ $y =$	$5n - 3 = 12$ $5n =$ $n =$	$3m - 7 = 14$ $3m =$ $m =$
$15 = 3a - 6$ $= 3a$ $= a$	$24 = 2w + 6$ $= 2w$ $= w$	$-2x + 9 = 23$ $-2x =$ $x =$	$20 - 3y = 2$ $-3y =$ $y =$
$\frac{1}{2}m + 6 = 18$ $\frac{1}{2}m =$ $m =$	$\frac{3}{4}n - 12 = 12$ $\frac{3}{4}n =$ $n =$	$3y + 1.5 = 6$ $3y =$ $y =$	$0.5w - 1.5 = 4.5$ $0.5w =$ $w =$

Mental Math

a.	**b.**	**c.**	**d.**
e.	**f.**	**g.**	**h.**

Problem Solving

Understand
What information am I given?
What am I asked to find or do?

Plan
How can I use the information I am given?
Which strategy should I try?

Solve
Did I follow the plan?
Did I show my work?
Did I write the answer?

Check
Did I use the correct information?
Did I do what was asked?
Is my answer reasonable?

Name ______________________ Time ____________

Power Up T

Use with Lesson 99

Facts Simplify.

$6 + 6 \times 6 - 6 \div 6 =$	$3^2 + \sqrt{4} + 5(6) - 7 + 8 =$
$4 + 2(3 + 5) - 6 \div 2 =$	$2 + 2[3 + 4(7 - 5)] =$
$\sqrt{1^3 + 2^3 + 3^3} =$	$\dfrac{4 + 3(7 - 5)}{6 - (5 - 4)} =$
$(-3)(-3) + (-3) - (-3) =$	$\dfrac{3(-3) - (-3)(-3)}{(-3) - (3)(-3)} =$

Mental Math

a.	**b.**	**c.**	**d.**
e.	**f.**	**g.**	**h.**

Problem Solving

Understand
What information am I given?
What am I asked to find or do?

Plan
How can I use the information I am given?
Which strategy should I try?

Solve
Did I follow the plan?
Did I show my work?
Did I write the answer?

Check
Did I use the correct information?
Did I do what was asked?
Is my answer reasonable?

Name ______________________ Time __________

Power Up U

Use with Lesson 100

Facts Complete each step to solve each equation.

$2x + 5 = 45$ $2x =$ $x =$	$3y + 4 = 22$ $3y =$ $y =$	$5n - 3 = 12$ $5n =$ $n =$	$3m - 7 = 14$ $3m =$ $m =$
$15 = 3a - 6$ $= 3a$ $= a$	$24 = 2w + 6$ $= 2w$ $= w$	$-2x + 9 = 23$ $-2x =$ $x =$	$20 - 3y = 2$ $-3y =$ $y =$
$\frac{1}{2}m + 6 = 18$ $\frac{1}{2}m =$ $m =$	$\frac{3}{4}n - 12 = 12$ $\frac{3}{4}n =$ $n =$	$3y + 1.5 = 6$ $3y =$ $y =$	$0.5w - 1.5 = 4.5$ $0.5w =$ $w =$

Mental Math

a.	b.	c.	d.
e.	f.	g.	h.

Problem Solving

Understand
What information am I given?
What am I asked to find or do?

Plan
How can I use the information I am given?
Which strategy should I try?

Solve
Did I follow the plan?
Did I show my work?
Did I write the answer?

Check
Did I use the correct information?
Did I do what was asked?
Is my answer reasonable?

Name ______________________ Time __________

Power Up Q

Use with Lesson 101

Facts Write the equivalent decimal and fraction for each percent.

Percent	Decimal	Fraction	Percent	Decimal	Fraction
10%			$33\frac{1}{3}\%$		
90%			20%		
5%			75%		
$12\frac{1}{2}\%$			$66\frac{2}{3}\%$		
50%			1%		
25%			250%		

Mental Math

a.	b.	c.	d.
e.	f.	g.	h.

Problem Solving

Understand

What information am I given?

What am I asked to find or do?

Plan

How can I use the information I am given?

Which strategy should I try?

Solve

Did I follow the plan?

Did I show my work?

Did I write the answer?

Check

Did I use the correct information?

Did I do what was asked?

Is my answer reasonable?

Name ____________________ Time __________

Power Up U

Use with Lesson 102

Facts

Complete each step to solve each equation.

$2x + 5 = 45$ $2x =$ $x =$	$3y + 4 = 22$ $3y =$ $y =$	$5n - 3 = 12$ $5n =$ $n =$	$3m - 7 = 14$ $3m =$ $m =$
$15 = 3a - 6$ $= 3a$ $= a$	$24 = 2w + 6$ $= 2w$ $= w$	$-2x + 9 = 23$ $-2x =$ $x =$	$20 - 3y = 2$ $-3y =$ $y =$
$\frac{1}{2}m + 6 = 18$ $\frac{1}{2}m =$ $m =$	$\frac{3}{4}n - 12 = 12$ $\frac{3}{4}n =$ $n =$	$3y + 1.5 = 6$ $3y =$ $y =$	$0.5w - 1.5 = 4.5$ $0.5w =$ $w =$

Mental Math

a.	**b.**	**c.**	**d.**
e.	**f.**	**g.**	**h.**

Problem Solving

Understand

What information am I given?

What am I asked to find or do?

Plan

How can I use the information I am given?

Which strategy should I try?

Solve

Did I follow the plan?

Did I show my work?

Did I write the answer?

Check

Did I use the correct information?

Did I do what was asked?

Is my answer reasonable?

Name ____________________ Time ____________

Power Up U

Use with Lesson 103

Facts Complete each step to solve each equation.

$2x + 5 = 45$ $2x =$ $x =$	$3y + 4 = 22$ $3y =$ $y =$	$5n - 3 = 12$ $5n =$ $n =$	$3m - 7 = 14$ $3m =$ $m =$
$15 = 3a - 6$ $= 3a$ $= a$	$24 = 2w + 6$ $= 2w$ $= w$	$-2x + 9 = 23$ $-2x =$ $x =$	$20 - 3y = 2$ $-3y =$ $y =$
$\frac{1}{2}m + 6 = 18$ $\frac{1}{2}m =$ $m =$	$\frac{3}{4}n - 12 = 12$ $\frac{3}{4}n =$ $n =$	$3y + 1.5 = 6$ $3y =$ $y =$	$0.5w - 1.5 = 4.5$ $0.5w =$ $w =$

Mental Math

a.	**b.**	**c.**	**d.**
e.	**f.**	**g.**	**h.**

Problem Solving

Understand
What information am I given?
What am I asked to find or do?

Plan
How can I use the information I am given?
Which strategy should I try?

Solve
Did I follow the plan?
Did I show my work?
Did I write the answer?

Check
Did I use the correct information?
Did I do what was asked?
Is my answer reasonable?

Name ______________________ Time __________

Power Up V

Use with Lesson 104

Facts Solve each equation.

$6x + 2x =$	$6x - 2x =$	$(6x)(2x) =$	$\frac{6x}{2x} =$
$9xy + 3xy =$	$9xy - 3xy =$	$(9xy)(3xy) =$	$\frac{9xy}{3xy} =$
$x + y + x =$	$x + y - x =$	$(x)(y)(-x) =$	$\frac{xy}{x} =$
$3x + x + 3 =$	$3x - x - 3 =$	$(3x)(-x)(-3) =$	$\frac{(2x)(8xy)}{4y} =$

Mental Math

a.	b.	c.	d.
e.	f.	g.	h.

Problem Solving

Understand

What information am I given?

What am I asked to find or do?

Plan

How can I use the information I am given?

Which strategy should I try?

Solve

Did I follow the plan?

Did I show my work?

Did I write the answer?

Check

Did I use the correct information?

Did I do what was asked?

Is my answer reasonable?

Name ______________________ Time __________

Power Up Q

Use with Lesson 105

Facts Write the equivalent decimal and fraction for each percent.

Percent	Decimal	Fraction	Percent	Decimal	Fraction
10%			$33\frac{1}{3}\%$		
90%			20%		
5%			75%		
$12\frac{1}{2}\%$			$66\frac{2}{3}\%$		
50%			1%		
25%			250%		

Mental Math

a.	**b.**	**c.**	**d.**
e.	**f.**	**g.**	**h.**

Problem Solving

Understand

What information am I given?
What am I asked to find or do?

Plan

How can I use the information I am given?
Which strategy should I try?

Solve

Did I follow the plan?
Did I show my work?
Did I write the answer?

Check

Did I use the correct information?
Did I do what was asked?
Is my answer reasonable?

Name ____________________ Time __________

Power Up V

Use with Lesson 106

Facts Solve each equation.

$6x + 2x =$	$6x - 2x =$	$(6x)(2x) =$	$\frac{6x}{2x} =$
$9xy + 3xy =$	$9xy - 3xy =$	$(9xy)(3xy) =$	$\frac{9xy}{3xy} =$
$x + y + x =$	$x + y - x =$	$(x)(y)(-x) =$	$\frac{xy}{x} =$
$3x + x + 3 =$	$3x - x - 3 =$	$(3x)(-x)(-3) =$	$\frac{(2x)(8xy)}{4y} =$

Mental Math

a.	**b.**	**c.**	**d.**
e.	**f.**	**g.**	**h.**

Problem Solving

Understand
What information am I given?
What am I asked to find or do?

Plan
How can I use the information I am given?
Which strategy should I try?

Solve
Did I follow the plan?
Did I show my work?
Did I write the answer?

Check
Did I use the correct information?
Did I do what was asked?
Is my answer reasonable?

Name ____________________ Time __________

Power Up Q

Use with Lesson 107

Facts

Write the equivalent decimal and fraction for each percent.

Percent	Decimal	Fraction	Percent	Decimal	Fraction
10%			$33\frac{1}{3}\%$		
90%			20%		
5%			75%		
$12\frac{1}{2}\%$			$66\frac{2}{3}\%$		
50%			1%		
25%			250%		

Mental Math

a.	**b.**	**c.**	**d.**
e.	**f.**	**g.**	**h.**

Problem Solving

Understand
What information am I given?
What am I asked to find or do?

Plan
How can I use the information I am given?
Which strategy should I try?

Solve
Did I follow the plan?
Did I show my work?
Did I write the answer?

Check
Did I use the correct information?
Did I do what was asked?
Is my answer reasonable?

Name ____________________ Time ____________

Power Up V

Use with Lesson 108

Facts

Solve each equation.

$6x + 2x =$	$6x - 2x =$	$(6x)(2x) =$	$\frac{6x}{2x} =$
$9xy + 3xy =$	$9xy - 3xy =$	$(9xy)(3xy) =$	$\frac{9xy}{3xy} =$
$x + y + x =$	$x + y - x =$	$(x)(y)(-x) =$	$\frac{xy}{x} =$
$3x + x + 3 =$	$3x - x - 3 =$	$(3x)(-x)(-3) =$	$\frac{(2x)(8xy)}{4y} =$

Mental Math

a.	**b.**	**c.**	**d.**
e.	**f.**	**g.**	**h.**

Problem Solving

Understand

What information am I given?

What am I asked to find or do?

Plan

How can I use the information I am given?

Which strategy should I try?

Solve

Did I follow the plan?

Did I show my work?

Did I write the answer?

Check

Did I use the correct information?

Did I do what was asked?

Is my answer reasonable?

Name ________________________ Time __________

Power Up V

Use with Lesson 109

Facts Solve each equation.

$6x + 2x =$	$6x - 2x =$	$(6x)(2x) =$	$\frac{6x}{2x} =$
$9xy + 3xy =$	$9xy - 3xy =$	$(9xy)(3xy) =$	$\frac{9xy}{3xy} =$
$x + y + x =$	$x + y - x =$	$(x)(y)(-x) =$	$\frac{xy}{x} =$
$3x + x + 3 =$	$3x - x - 3 =$	$(3x)(-x)(-3) =$	$\frac{(2x)(8xy)}{4y} =$

Mental Math

a.	**b.**	**c.**	**d.**
e.	**f.**	**g.**	**h.**

Problem Solving

Understand
What information am I given?
What am I asked to find or do?

Plan
How can I use the information I am given?
Which strategy should I try?

Solve
Did I follow the plan?
Did I show my work?
Did I write the answer?

Check
Did I use the correct information?
Did I do what was asked?
Is my answer reasonable?

Name ____________________ Time __________

Power Up Q

Use with Lesson 110

Facts Write the equivalent decimal and fraction for each percent.

Percent	Decimal	Fraction	Percent	Decimal	Fraction
10%			$33\frac{1}{3}\%$		
90%			20%		
5%			75%		
$12\frac{1}{2}\%$			$66\frac{2}{3}\%$		
50%			1%		
25%			250%		

Mental Math

a.	**b.**	**c.**	**d.**
e.	**f.**	**g.**	**h.**

Problem Solving

Understand
What information am I given?
What am I asked to find or do?

Plan
How can I use the information I am given?
Which strategy should I try?

Solve
Did I follow the plan?
Did I show my work?
Did I write the answer?

Check
Did I use the correct information?
Did I do what was asked?
Is my answer reasonable?

Name ______________________ Time ____________

Power Up V

Use with Lesson 111

Facts

Solve each equation.

$6x + 2x =$	$6x - 2x =$	$(6x)(2x) =$	$\frac{6x}{2x} =$
$9xy + 3xy =$	$9xy - 3xy =$	$(9xy)(3xy) =$	$\frac{9xy}{3xy} =$
$x + y + x =$	$x + y - x =$	$(x)(y)(-x) =$	$\frac{xy}{x} =$
$3x + x + 3 =$	$3x - x - 3 =$	$(3x)(-x)(-3) =$	$\frac{(2x)(8xy)}{4y} =$

Mental Math

a.	**b.**	**c.**	**d.**
e.	**f.**	**g.**	**h.**

Problem Solving

Understand

What information am I given?

What am I asked to find or do?

Plan

How can I use the information I am given?

Which strategy should I try?

Solve

Did I follow the plan?

Did I show my work?

Did I write the answer?

Check

Did I use the correct information?

Did I do what was asked?

Is my answer reasonable?

Name ____________________ Time __________

Power Up W

Use with Lesson 112

Facts

Simplify. Write each answer in scientific notation.

$(1 \times 10^6)(1 \times 10^6) =$	$(3 \times 10^3)(3 \times 10^3) =$	$(4 \times 10^{-5})(2 \times 10^{-6}) =$
$(5 \times 10^5)(5 \times 10^5) =$	$(6 \times 10^{-3})(7 \times 10^{-4}) =$	$(3 \times 10^6)(2 \times 10^{-4}) =$
$\frac{8 \times 10^8}{2 \times 10^2} =$	$\frac{5 \times 10^6}{2 \times 10^3} =$	$\frac{9 \times 10^3}{3 \times 10^8} =$
$\frac{2 \times 10^6}{4 \times 10^2} =$	$\frac{1 \times 10^{-3}}{4 \times 10^8} =$	$\frac{8 \times 10^{-8}}{2 \times 10^{-2}} =$

Mental Math

a.	b.	c.	d.
e.	f.	g.	h.

Problem Solving

Understand
What information am I given?
What am I asked to find or do?

Plan
How can I use the information I am given?
Which strategy should I try?

Solve
Did I follow the plan?
Did I show my work?
Did I write the answer?

Check
Did I use the correct information?
Did I do what was asked?
Is my answer reasonable?

Name ______________________ Time __________

Power Up V

Use with Lesson 113

Facts Solve each equation.

$6x + 2x =$	$6x - 2x =$	$(6x)(2x) =$	$\frac{6x}{2x} =$
$9xy + 3xy =$	$9xy - 3xy =$	$(9xy)(3xy) =$	$\frac{9xy}{3xy} =$
$x + y + x =$	$x + y - x =$	$(x)(y)(-x) =$	$\frac{xy}{x} =$
$3x + x + 3 =$	$3x - x - 3 =$	$(3x)(-x)(-3) =$	$\frac{(2x)(8xy)}{4y} =$

Mental Math

a.	**b.**	**c.**	**d.**
e.	**f.**	**g.**	**h.**

Problem Solving

Understand
What information am I given?
What am I asked to find or do?

Plan
How can I use the information I am given?
Which strategy should I try?

Solve
Did I follow the plan?
Did I show my work?
Did I write the answer?

Check
Did I use the correct information?
Did I do what was asked?
Is my answer reasonable?

Name ______________________ Time __________

Power Up V

Use with Lesson 114

Facts

Solve each equation.

$6x + 2x =$	$6x - 2x =$	$(6x)(2x) =$	$\frac{6x}{2x} =$
$9xy + 3xy =$	$9xy - 3xy =$	$(9xy)(3xy) =$	$\frac{9xy}{3xy} =$
$x + y + x =$	$x + y - x =$	$(x)(y)(-x) =$	$\frac{xy}{x} =$
$3x + x + 3 =$	$3x - x - 3 =$	$(3x)(-x)(-3) =$	$\frac{(2x)(8xy)}{4y} =$

Mental Math

a.	**b.**	**c.**	**d.**
e.	**f.**	**g.**	**h.**

Problem Solving

Understand

What information am I given?

What am I asked to find or do?

Plan

How can I use the information I am given?

Which strategy should I try?

Solve

Did I follow the plan?

Did I show my work?

Did I write the answer?

Check

Did I use the correct information?

Did I do what was asked?

Is my answer reasonable?

Name ______________________ Time __________

Power Up V

Use with Lesson 115

Facts

Solve each equation.

$6x + 2x =$	$6x - 2x =$	$(6x)(2x) =$	$\frac{6x}{2x} =$
$9xy + 3xy =$	$9xy - 3xy =$	$(9xy)(3xy) =$	$\frac{9xy}{3xy} =$
$x + y + x =$	$x + y - x =$	$(x)(y)(-x) =$	$\frac{xy}{x} =$
$3x + x + 3 =$	$3x - x - 3 =$	$(3x)(-x)(-3) =$	$\frac{(2x)(8xy)}{4y} =$

Mental Math

a.	**b.**	**c.**	**d.**
e.	**f.**	**g.**	**h.**

Problem Solving

Understand

What information am I given?

What am I asked to find or do?

Plan

How can I use the information I am given?

Which strategy should I try?

Solve

Did I follow the plan?

Did I show my work?

Did I write the answer?

Check

Did I use the correct information?

Did I do what was asked?

Is my answer reasonable?

Name ______________________ Time ____________

Power Up W

Use with Lesson 116

Facts

Simplify. Write each answer in scientific notation.

$(1 \times 10^6)(1 \times 10^6) =$	$(3 \times 10^3)(3 \times 10^3) =$	$(4 \times 10^{-5})(2 \times 10^{-6}) =$
$(5 \times 10^5)(5 \times 10^5) =$	$(6 \times 10^{-3})(7 \times 10^{-4}) =$	$(3 \times 10^6)(2 \times 10^{-4}) =$
$\frac{8 \times 10^8}{2 \times 10^2} =$	$\frac{5 \times 10^6}{2 \times 10^3} =$	$\frac{9 \times 10^3}{3 \times 10^8} =$
$\frac{2 \times 10^6}{4 \times 10^2} =$	$\frac{1 \times 10^{-3}}{4 \times 10^8} =$	$\frac{8 \times 10^{-8}}{2 \times 10^{-2}} =$

Mental Math

a.	**b.**	**c.**	**d.**
e.	**f.**	**g.**	**h.**

Problem Solving

Understand

What information am I given?

What am I asked to find or do?

Plan

How can I use the information I am given?

Which strategy should I try?

Solve

Did I follow the plan?

Did I show my work?

Did I write the answer?

Check

Did I use the correct information?

Did I do what was asked?

Is my answer reasonable?

Name ______________________ Time __________

Power Up W

Use with Lesson 117

Facts Simplify. Write each answer in scientific notation.

$(1 \times 10^6)(1 \times 10^6) =$	$(3 \times 10^3)(3 \times 10^3) =$	$(4 \times 10^{-5})(2 \times 10^{-6}) =$
$(5 \times 10^5)(5 \times 10^5) =$	$(6 \times 10^{-3})(7 \times 10^{-4}) =$	$(3 \times 10^6)(2 \times 10^{-4}) =$
$\frac{8 \times 10^8}{2 \times 10^2} =$	$\frac{5 \times 10^6}{2 \times 10^3} =$	$\frac{9 \times 10^3}{3 \times 10^8} =$
$\frac{2 \times 10^6}{4 \times 10^2} =$	$\frac{1 \times 10^{-3}}{4 \times 10^8} =$	$\frac{8 \times 10^{-8}}{2 \times 10^{-2}} =$

Mental Math

a.	**b.**	**c.**	**d.**
e.	**f.**	**g.**	**h.**

Problem Solving

Understand
What information am I given?
What am I asked to find or do?

Plan
How can I use the information I am given?
Which strategy should I try?

Solve
Did I follow the plan?
Did I show my work?
Did I write the answer?

Check
Did I use the correct information?
Did I do what was asked?
Is my answer reasonable?

Name ______________________ Time __________

Power Up W

Use with Lesson 118

Facts Simplify. Write each answer in scientific notation.

$(1 \times 10^6)(1 \times 10^6) =$	$(3 \times 10^3)(3 \times 10^3) =$	$(4 \times 10^{-5})(2 \times 10^{-6}) =$
$(5 \times 10^5)(5 \times 10^5) =$	$(6 \times 10^{-3})(7 \times 10^{-4}) =$	$(3 \times 10^6)(2 \times 10^{-4}) =$
$\frac{8 \times 10^8}{2 \times 10^2} =$	$\frac{5 \times 10^6}{2 \times 10^3} =$	$\frac{9 \times 10^3}{3 \times 10^8} =$
$\frac{2 \times 10^6}{4 \times 10^2} =$	$\frac{1 \times 10^{-3}}{4 \times 10^8} =$	$\frac{8 \times 10^{-8}}{2 \times 10^{-2}} =$

Mental Math

a.	**b.**	**c.**	**d.**
e.	**f.**	**g.**	**h.**

Problem Solving

Understand
What information am I given?
What am I asked to find or do?

Plan
How can I use the information I am given?
Which strategy should I try?

Solve
Did I follow the plan?
Did I show my work?
Did I write the answer?

Check
Did I use the correct information?
Did I do what was asked?
Is my answer reasonable?

Name ______________________ Time __________

Power Up V

Use with Lesson 119

Facts Solve each equation.

$6x + 2x =$	$6x - 2x =$	$(6x)(2x) =$	$\frac{6x}{2x} =$
$9xy + 3xy =$	$9xy - 3xy =$	$(9xy)(3xy) =$	$\frac{9xy}{3xy} =$
$x + y + x =$	$x + y - x =$	$(x)(y)(-x) =$	$\frac{xy}{x} =$
$3x + x + 3 =$	$3x - x - 3 =$	$(3x)(-x)(-3) =$	$\frac{(2x)(8xy)}{4y} =$

Mental Math

a.	**b.**	**c.**	**d.**
e.	**f.**	**g.**	**h.**

Problem Solving

Understand

What information am I given?

What am I asked to find or do?

Plan

How can I use the information I am given?

Which strategy should I try?

Solve

Did I follow the plan?

Did I show my work?

Did I write the answer?

Check

Did I use the correct information?

Did I do what was asked?

Is my answer reasonable?

Name ______________________ Time ____________

Power Up W

Use with Lesson 120

Facts Simplify. Write each answer in scientific notation.

$(1 \times 10^6)(1 \times 10^6) =$	$(3 \times 10^3)(3 \times 10^3) =$	$(4 \times 10^{-5})(2 \times 10^{-6}) =$
$(5 \times 10^5)(5 \times 10^5) =$	$(6 \times 10^{-3})(7 \times 10^{-4}) =$	$(3 \times 10^6)(2 \times 10^{-4}) =$
$\frac{8 \times 10^8}{2 \times 10^2} =$	$\frac{5 \times 10^6}{2 \times 10^3} =$	$\frac{9 \times 10^3}{3 \times 10^8} =$
$\frac{2 \times 10^6}{4 \times 10^2} =$	$\frac{1 \times 10^{-3}}{4 \times 10^8} =$	$\frac{8 \times 10^{-8}}{2 \times 10^{-2}} =$

Mental Math

a.	**b.**	**c.**	**d.**
e.	**f.**	**g.**	**h.**

Problem Solving

Understand
What information am I given?
What am I asked to find or do?

Plan
How can I use the information I am given?
Which strategy should I try?

Solve
Did I follow the plan?
Did I show my work?
Did I write the answer?

Check
Did I use the correct information?
Did I do what was asked?
Is my answer reasonable?

